Pond Watching

Pond Watching

Paul Sterry

With drawings by the author

SEVERN HOUSE NATURALIST'S LIBRARY

To my family and friends with thanks for their encouragement

British Library Cataloguing in Publication Data

Sterry, Paul
Pond Watching.
– (Severn House naturalist's library)
1. Fresh-water biology
I. Title
574.92'9 QH90
ISBN 0-7278-2025-7

Published by Severn House Publishers Limited
4 Brook Street
London W1Y 1AA

Editorial Ian Jackson and Diana Levinson

Typeset by DP Press Limited, Sevenoaks, Kent

Printed and bound by Hazell Watson and Viney Ltd. Aylesbury, Buckinghamshire

Contents

1 The freshwater environment

Freshwater is a haven for wildlife and consequently for the naturalist. Even the water's surface teems with life and many useful observations can be made whilst peacefully sitting at a pond's edge. Below the surface there can be an abundance of life as well, providing a never-ending source of delight for the pond hunter. There can be few people who have not dipped a net into a pond at some point in their lives. Ponds are ideal spots to nurture an enthusiasm for natural history especially amongst children and, as with most hobbies, you will find that the more you learn, the more you will want to discover.

This book concentrates on ponds rather than other freshwater habitats; they are invariably more productive than lakes and rivers both in terms of numbers of animals present and species diversity, and for the amateur naturalist, they are easier to study. Ponds are rather difficult to define, but for our purposes they can be considered as small bodies of freshwater, shallow enough for vegetation to grow across the whole surface area. There are obvious exceptions, with small, deep pools and shallow lakes being difficult to categorise. The description of 'pond' life will also apply to a wide variety of other types of freshwater habitat such as canals, gravel pits and lowland lakes. It is useful to compare ponds with these other water bodies and this is one of the projects discussed in Chapter 6.

The study of freshwater life is not simply a question of identification. This is an essential and fascinating part of pond hunting and some people find it sufficient to maintain their interest. It is perhaps more rewarding though to use identification as a tool to enable you to study a pond in detail through the seasons and also to compare it with others.

The properties of water

The first step is to be aware of some of the special properties of water itself. This is helpful in explaining many of the adaptations of freshwater animals and plants and the factors influencing their distribution, not only within an individual pond but also on a national basis.

The solubility of gases

Gases will dissolve in water and the levels of the two most important – oxygen and carbon dioxide – are critical to freshwater life. Rainwater is basically pure, but during its travels it picks up small amounts of these gases. Much more carbon dioxide and oxygen become dissolved when the water is on the ground as I will describe.

Oxygen is essential to respiration in animals and plants. It diffuses into water from the atmosphere and is supplemented by photosynthesis in aquatic plants. (This is the conversion of carbon dioxide and water to glucose and oxygen using sunlight energy.) Air contains approximately 210 ml oxygen per litre, about 30 times the level found in freshwater. So there is a continual supply of aerial oxygen, but the amount that a given volume of water can dissolve by diffusion alone is constant. If this level is reached then the water becomes 'saturated' with oxygen. However, the level found in nature is not always the saturation value. It depends in part on the respiratory requirements of the animals and plants in the water. The greater the biomass, the more oxygen will be used up. During daylight hours, plants photosynthesise and produce oxygen which may elevate the dissolved oxygen levels to a point known as 'supersaturation'. Also, a flowing water system will have higher oxygen levels than a standing water body because turbulence will cause aeration. In general, a freshwater habitat where the water is always saturated or supersaturated with oxygen can support more forms of life than an area where this is not the case. Indeed, there are many animals, such as mayfly nymphs, which breathe dissolved oxygen and will only survive in well aerated water. So the oxygen level of a water body can have a dramatic effect on the life that it holds.

To complicate matters further, the saturation values for water are not absolutely constant but vary with temperature: the higher the temperature, the less oxygen can dissolve in the water. Water can hold about 9 ml oxygen per litre at 4°C, but only 7 ml per litre at 15°C. This provides a problem for freshwater animals for, with an increase in temperature, not only will less oxygen be available for them, but they will actually require more because their metabolism increases. Several factors help to overcome this difficulty. Many freshwater animals such as insects are only secondarily adapted to water. Some, such as the larva of the great diving beetle, breathe aerial oxygen directly, whilst others, like the adult beetle, carry a bubble of air around with them. The truly aquatic animals have gills which increase the surface area available for gas exchange. As this surface area to volume ratio determines whether an animal can depend on simple diffusion for its oxygen requirements, only the

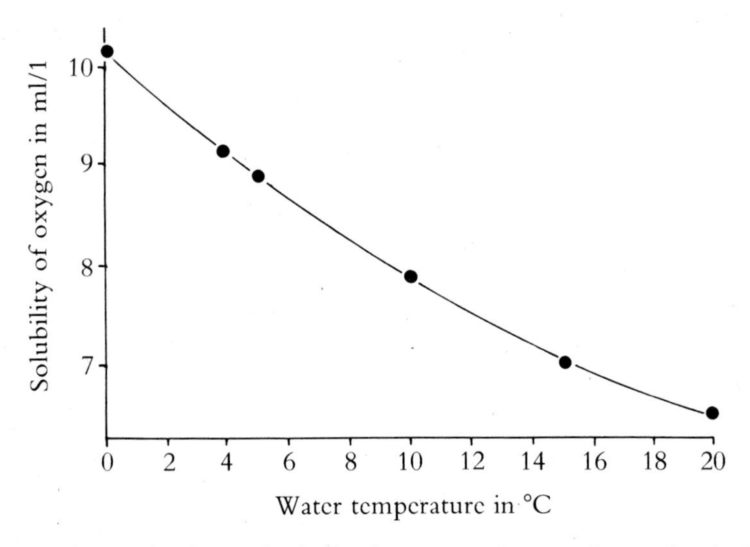

Figure 1 This graph indicates clearly that the amount of oxygen that can dissolve in water decreases as the temperature increases. Freshwater animals have various adaptations which help them cope with this.

smallest of pond creatures with low metabolic rates, such as protozoa and planktonic crustacea can rely on this method.

Oxygen levels also vary diurnally. Again, temperature will be an important factor, as it is generally warmer during the day than at night. The fluctuations may also be related to the plants and animals. During the day, photosynthesis may result in supersaturation, while at night both plants and animals will be respiring and using up this oxygen.

Decay processes use up a lot of oxygen. As a consequence, the bottom silt in a pond where they take place may be totally anaerobic (oxygen free). Many mud-dwelling animals therefore use water from the water column above which will contain oxygen. Organic pollution may encourage the growth of bacteria and fungi. These are the major decay organisms and a vast increase in their numbers may seriously deplete oxygen levels, effectively 'killing' the water.

Carbon dioxide is one of the end-products of respiration but also one of the prerequisites of photosynthesis. As such, it is an extremely important gas. Carbon dioxide is soluble in water and during its travels, rainwater acquires a small amount. This forms a weak solution of carbonic acid. Greater concentrations are acquired by the water body itself. It comes mainly from the air by diffusion, but this is supplemented by the respiration of aquatic life.

The solubility of salts

Carbon dioxide, as well as existing in its dissolved form, can also produce carbonate and bicarbonate salts of, in particular, calcium and magnesium. In chalk areas, where there is plenty of calcium, a considerable amount of carbon dioxide is lost from the water to form calcium carbonate. This salt-rich, alkaline water is termed hard. Molluscs and crustaceans which incorporate calcium into their shells and exoskeletons enjoy a hard water environment. Hard waters are therefore generally more productive than soft waters which contain few dissolved salts. If water has peat origins, then it is acidic and will have a low productivity rate. Fewer bacteria can survive which means that decay is slow and hence less nutrients are recycled. Because of this plant growth is poor which adversely affects animal numbers. (This is discussed in more detail in Chapter 6.)

As with oxygen, the level of carbon dioxide and its salts is subject to daily and seasonal changes. During the day photosynthesis uses up carbon dioxide, while at night both animals and plants produce it through respiration. Seasonal changes affect both temperature and the life of aquatic plants. In the winter many plants, in particular the emergent species, die back. Photosynthesis occurs at a lower rate than in the summer because of this and less carbon dioxide is used up. More decay takes place in autumn and winter thus releasing more carbon dioxide. The overall levels of carbon dioxide are therefore higher in the winter.

The quantity and type of salts dissolved in water depends largely on the nature of the bedrock of the surrounding area. We have already seen how a chalk soil can contribute to the hardness of the water. There are many other chemical constituents in the water, such as nitrates and sulphates, and their levels, in part, determine the flora and fauna found there. Salts are also recycled by the processes of decomposition, and this too can be very important for plant and animal growth.

Like crustaceans and molluscs, diatoms have a hard outer casing, but instead of calcium, they use silica to strengthen their shells. Consequently they are present only in water which contains it and can actually cause a noticeable reduction in the silica levels illustrating what ecologists term 'resource depletion'. Their numbers increase rapidly while silica is present in the water. Eventually they use it all up and the population dies back until the following year.

The amount of dissolved sodium chloride (table salt) is important for freshwater life. Freshwater always contains a lower concentration than that in the cells of pond organisms. Their bodies are permeable to water which continually enters their cells to equalise this difference in concentrations, (osmosis). Pond animals would be constantly swelling up if they did not remove the water as fast as it entered them. Single-celled animals such as *Paramecium* have contractile vacuoles which remove excess water. More advanced animals have kidneys which perform this function and also remove nitrogenous waste. In the sea, the salt concentration exceeds that inside animals and so marine animals have the reverse problem – they must try to retain water.

Osmosis or water regulation can be easily demonstrated using uncooked potato chips. If they are left in tapwater and in strong salt solution for half an hour, a difference in their appearance will be observed. The chips in tapwater will have become swollen and hard because all the cells will be 'turgid' or full of water while those in salt solution will have lost a lot of water and will be limp.

The absorption of light

Light passes through air without much loss in intensity. Water is a denser medium and consequently absorbs more light. In addition, water contains particles of sediment and the degree of turbidity these produce can have dramatic effects on light penetration. Plants need light for photosynthesis and so seldom grow below about 10 m in a lake; there is simply not enough light below that depth. Many aquatic plants such as water lilies overcome this problem by having floating leaves. Others, such as the duckweeds, float entirely on the surface.

Temperature

Water is a poor conductor of heat and this has profound effects upon freshwater life. Biological reactions, like chemical ones, work at faster rates as the temperature increases. Since aquatic invertebrates have body temperatures identical to the surrounding water, they are at its mercy.

Water has the unique property of being most dense as a liquid and not as a solid at 4°C. The bottom waters in a lake remain at 4°C even during the summer forming a distinct layer called the hypolimnion. From the top to the bottom of the hypolimnion there will be little change in temperature. At the surface, another layer of water, the epilimnion, is warmed by the sun and air. Across this layer there will also be little change in temperature. Between the hypolimnion and epilimnion there is a zone of mixing with a sharp temperature gradient, called the thermocline. The depth of this and the epilimnion depends upon the strength of the sun's rays and the degree of mixing caused by wind action.

In a small pond during the summer, the sun will warm the surface waters and the wind will cause mixing. If the pond is shallow enough there will be only a small temperature gradient between top and bottom during the day. At night, heat will be lost by convection to the air from the upper layers and so a temperature gradient may form. There is often a marked daily temperature fluctuation, with

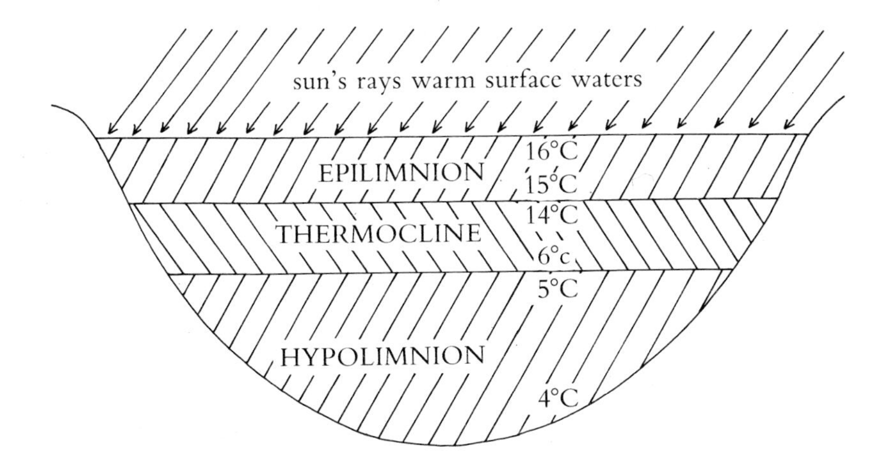

Figure 2 This cross-section through a pond shows the division into epilimnion, thermocline and hypolimnion with the thermocline having the most marked temperature gradient.

the water temperature following changes in air temperature. The delay or lag between the air and water temperatures depends upon the size of the pond. Temperature is one factor affecting vertical zonation in the water, but most pond organisms can cope with some variation.

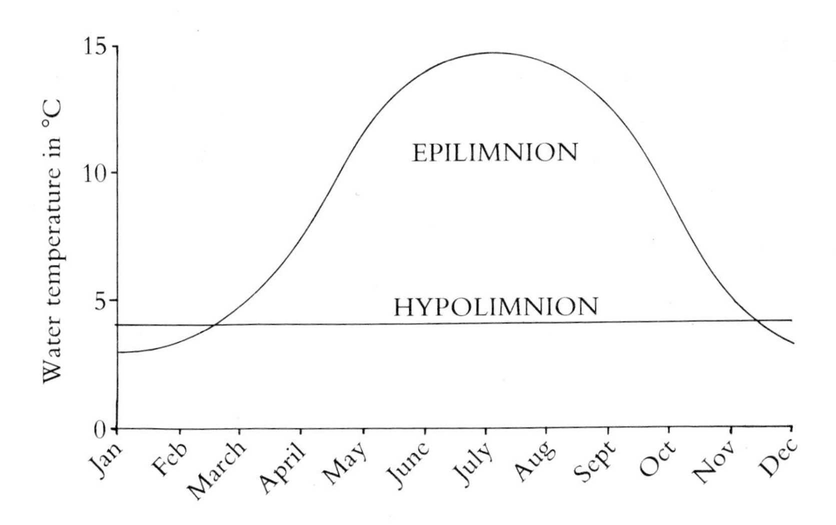

Figure 3 In the summer months the temperature in the epilimnion can rise as high as 15°C, falling in the autumn and spring to about 4°C. In contrast, the hypolimnion maintains a constant temperature throughout the year of between 4° and 5°C.

During the autumn, the temperature of a lake will gradually drop with the air temperature. As the surface water loses heat it becomes more dense and sinks. The warmer water below will rise and is subsequently cooled. Soon the whole water body may reach 4°C. Below this temperature the colder water floats and the warmer water sinks, which is why ice forms on the surface first. Ice is an even poorer conductor of heat than water and since heat loss can only occur through the ice after it has formed, the lake is effectively insulated. Only in very severe conditions does the ice layer become thick, and consequently freshwater life in Britain seldom experiences freezing temperatures.

The density of water

The density of water changes with temperature, but it is always far greater than air. This gives freshwater life buoyancy and the organisms are often far less rigid than their terrestrial counterparts. The bodies of smaller invertebrates have a density similar to the water and therefore have freedom of movement. Water plants too are supported by the water and in many cases cannot exist on land. Larger animals are denser than water but many gain more buoyancy by carrying air around with them. Generally they need only a small volume to counteract gravity. Fish have a specialised swim bladder and regulate the volume of air it contains so that they can either float or sink. Some aquatic snails carry air within their shells which makes them less cumbersome. It also enables them to escape predators because by expelling the air they can sink rapidly to the bottom. Many insects such as water bugs and beetles carry an air supply with them for both buoyancy and respiration. In some species such as the water boatman their air supply causes them to bob to the surface; to stay down, they have to hold on to something.

Figure 4 Many animals like the water boatman live at the surface film. The boatman detects vibrations in the surface produced by drowning insects on which it feeds.

The surface film

The surface film has the curious property of clinging tightly to itself, a phenomenon known as surface tension. Many animals, such as pond skaters exploit this property and increase their ability to remain on the surface film by having a wetproof body. Waxy hairs and secretions are extremely water-repellent and are used by most surface animals in a way analagous to the resistance of a newly waxed car to rain. The water forms large droplets instead of an even film because the molecules have a greater affinity for each other than for the wax polish. Detergent in the water can severely reduce surface tension and this form of pollution sometimes occurs when factory effluent is pumped into a river.

I hope that you now have some understanding of how the adaptations and distribution of freshwater organisms relate to the water itself. They are not only dependent on its special properties but also influenced by the environment around it.

2 Freshwater habitats

There is a tremendous wealth and variety of freshwater habitat in Britain. The origins of these water bodies are as diverse as the habitats themselves and an understanding of these origins is helpful in knowing what to look for in them.

The source of the water that fills our rivers, ponds and lakes is rain. How the rain reaches the water body is important because the gases and salts it absorbs on the way influence the flora and fauna.

In most of lowland Britain the soil is porous and immediately absorbs any rain falling on it. The water filters down through the soil to the bedrock and on its way dissolves salts and nutrients which alter its composition. Eventually, springs give rise to streams which make their progress to the sea. In heathland areas where peat builds up, the soil is often so saturated with water that rain remains on the surface and trickles off. Here the water will have less opportunity to dissolve nutrients and so supports fewer plants and animals. In upland regions where the soil or rock is often not porous the water runs straight off into streams. Again, the water will be able to dissolve fewer nutrients than in lowland regions.

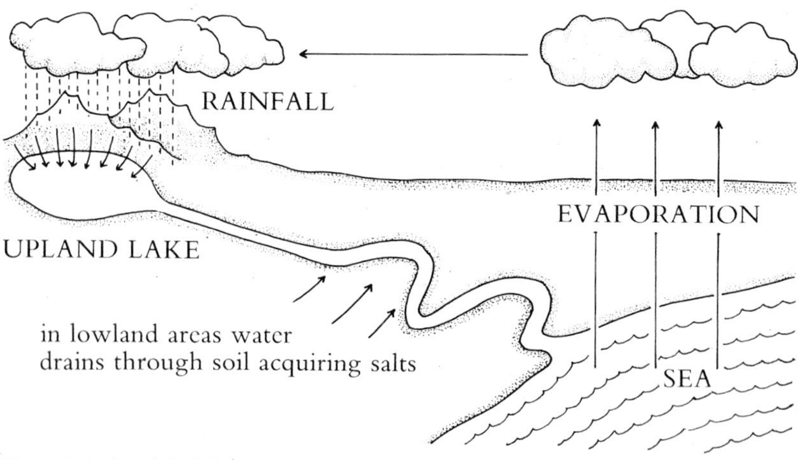

Figure 5 A simplified diagram of the water cycle.

In small ponds the water level may be regulated entirely by rain falling into it and water evaporating. However, in larger water bodies such as lakes, streams may flow into and out of the system. The size of a water body, its geographical position and its current use by Man all influence the life within it.

Natural habitats

Streams and rivers

Most of our streams and rivers are of natural origin, even though their courses may have been altered by Man. Although their sources may be diverse, all rivers can be divided into four regions.

Near the source of an upland stream the current will be fast-flowing. A more significant factor for freshwater life, is the bottom substrate which moves continually. This part of the river's course is called the headstream and may be given to fluctuations in temperature and volume. As a consequence it supports little life.

Below this is the troutbeck region. The current here is still torrential but its volume is more constant. The river will now be larger and the substrate more stable, consisting of boulders with gravel in between. The troutbeck is so-called because it is a haven for trout which feed on the mayfly and stonefly nymphs which live there. Although there may only be a few species of insects living in the troutbeck, their populations can be large.

As the river moves to lower altitudes, the speed of the current decreases. The volume of water carried by this region, called the minnow-reach, is greater because subsidiary streams have joined

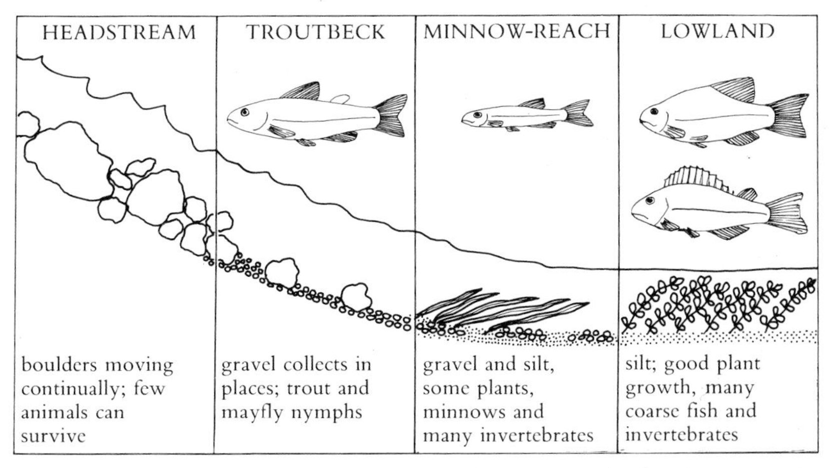

Figure 6 The four major regions of a river: headstream, troutbeck, minnow-reach and lowland. As the force of the current decreases and the bedrock becomes more stable, a greater abundance of animal and plant life can flourish.

the main flow. The river will have acquired more nutrients by this stage and the bottom is even more stable. In some parts, pools may be created and silt may gather in the slower moving areas. This allows for a greater species diversity and the minnow-reach is the first region where plants can successfully colonise the water and the water's edge in any number.

After the minnow-reach, the river approaches the coast. The speed of the current will have decreased but the volume of the river will have increased even more. This area is usually the most productive part of the river and large fish populations will be present. These would not survive were it not for the extensive range of invertebrates also found here together with healthy plant communities.

Lakes and ponds

Most of the natural large water bodies in Britain are found in upland areas. Glaciation in past Ice Ages produced large U-shaped valleys which frequently collected water and formed lakes. Occasionally the terminal moraines of the glaciers may have blocked off valleys allowing large volumes of water to accumulate. Rainfall is generally high in upland areas and because the rock is often impervious, water is plentiful. On a lesser scale there are isolated tarns, also the result of glaciation, which contain much smaller volumes of water. These are generally at higher altitudes than the lakes and even less productive. The animals and plants found in upland regions are often specially adapted to this harsh environment, so much so that they are not usually found elsewhere.

It is in lowland areas that you will first come across natural freshwater habitats that are really rewarding to study. The vast majority of ponds and lakes in lowland areas are man-made, but in heathland areas, any pools are likely to be of natural origin. *Sphagnum* moss is one of the most typical heathland plants, forming a green carpet wherever there is water. When it dies, it decays slowly and layers build up on the peat deposits already present. This occurs most frequently in valley bottoms. Since peat and dead and living *Sphagnum* absorb water readily, the valley bottoms are often saturated throughout the year and it is here that pools and bogs develop. Pools are often created and fed by streams. The acidic nature of the water and its lack of nutrients make for a low bacterial content and turnround rate of organic material. As a consequence these waters support only a few species. They are particularly rich in dragonfly nymphs and water bugs, as there are fewer competing species, but contain hardly any species of snails or crustaceans

Figure 7 The banks of this typical stream are covered in vegetation and the water contains a wealth of plant and animal life.

Figure 10 A peaty pool in Cwm Idwal, Snowdonia with *Sphagnum* moss, cotton grass and rushes growing around the edge. Such water can support few animals because of its acidic content.

because of the lack of calcium.

As a river approaches the coast, the speed of the current decreases. If, in its final reaches, it flows over flat plains, it will tend to meander which sometimes results in the formation of pools called oxbow lakes. Here the water which has not followed the curves of the river simply cuts off a bend. The redundant bend then has silt deposited across its entrance which eventually cuts it off completely from the river.

Figure 11 Lowland rivers such as the Cuckmere in Sussex often meander and oxbow lakes form when the river's course cuts off a loop.

The ultimate fate of river water is to enter the sea. The sea may carry silt or wash larger particles of rock along the sea floor and the current of the river can cause some of this material to be deposited in the form of mud-bars or shingle ridges. On occasions the shingle

Figure 8 The marsh marigold is a typical plant of pond margins. Like the water crowfoot it is a member of the buttercup family.
Figure 9 There are many species of water crowfoot in Britain. The thread-leaved water crowfoot shown here is found in streams.

may cut off large areas of water which then remain predominantly freshwater as at Slapton Ley in Devon and Little Sea in Dorset, two of the few large, natural bodies of freshwater in lowland Britain. There is some similarity between these areas and the man-made lagoons found behind sea walls around our coasts. However the latter hardly fall into the category of freshwater habitats since the sea often greatly increases their salinity. Although some freshwater animals such as sticklebacks can tolerate a certain degree of salt, these areas are largely the domain of estuarine animals.

Not all natural freshwater habitats are large or have such a degree of permanence as those described. At the other end of the scale are the thousands of sizeable puddles and cart-ruts that regularly fill up and support life. In upland areas of Wales, I have seen places where nearly every available rut and puddle has a mass of frogspawn in it. In such situations a few tadpoles may survive, feeding at first on microscopic algae and later on each other. In lowland areas, puddles often have a good supply of microscopic protozoa and even a few crustaceans or mosquito larvae. Larger temporary pools may harbour the fairy shrimp and *Triops cancriformis*, two species of crustaceans adapted to this temporary environment.

Man-made habitats

Undoubtedly the majority of freshwater bodies that will be encountered by the pond hunter will be man-made or at least altered in some way. The most typical habitat is the pond, and for historical reasons this is invariably close to Man's dwellings.

The village pond

The pond used to be a focus of village life because of its essential role in providing water and, secondarily, for recreation. Ponds are often sited at the meeting point of several roads in the centre of the village. In the past, animals would have been driven along these roads to and from markets and pastures and carrying loads and the site of a pond was vitally important for watering them. In some cases the village may have developed after and even because of the pond as the people driving the animals would have provided the custom for the various trades in the village.

Where farms had large numbers of animals, a pond would have been a prerequisite and few long-established farms are without one. Where crops took precedence over stock, a pond was often sited at the junction of several fields and its water which was needed in times of drought, was carried to the fields by hand – a most laborious process.

On the downland of southern England standing water is often at a premium since it drains easily through the chalk. When sheep grazing became popular, some form of water supply was necessary

Figure 12 A typical village pond at Oakley in Hampshire sited in the centre of the village at the junction of several roads.

and dewponds were created. These were shallow circular depressions, lined with clay. The clay was well puddled and it contained flints to lessen the damage caused by animals' feet. In country lore dewponds often have a certain mystique, but their origins are more down to earth; catching rainwater is more important than catching dew despite their popular name. Although some dewponds are thought to be ancient, the oldest still in use probably date back no earlier than the seventeenth century.

In medieval times, wetland areas were also used to grow sallow and the withy beds were harvested to produce long thin wands which were used in trades as diverse as basket weaving and house-building. Wetlands are still important today for the growth of the reed *Phragmites communis*, the major component in thatching which was used far more extensively in the past.

Figure 13 The traditional craft of thatching still survives. Here osiers are stacked for drying at Sedgemoor.

Some village ponds were probably stocked with fish to supplement the diet of the local population. However, with all the trampling and disturbance, they may not have lasted well. In past years all our rivers would have had good fish populations although these would not always have been easy to catch. In medieval times and even earlier, monks and landowners had the better idea of creating stewponds which were stocked with fish such as carp and tench so that there was a ready supply at hand. Moats around castles were also obvious sites for stewponds. In times of siege the occupants would have had a source of fresh fish – if they were willing to dodge the arrows!

Stewponds would have been more than just a luxury in some areas because of the lack of alternative animal protein. During the winter months a large proportion of the livestock would have been slaughtered since there would have been little to feed them on. Fish can feed themselves and could be eaten in times of need. Also the Church decreed that fish should be eaten on Fridays and during Lent and this additional religious requirement would have encouraged the upkeep of stewponds.

Millponds

At the onset of the Industrial Revolution, the village pond was falling into decline. It could not meet the increased industrial demand for water and so, continued to disappear. Other types of water body, however, found new and more profitable uses. One example was the millpond. These were created in feudal times to overcome fluctuations in the water supply and provide a good head of water upstream to power the mill's grind stone. Millponds were later used in small numbers to power iron foundries, but the

Figure 14 A water mill and wheel at Lower Slaughter, Gloucestershire. Water mills were widely used during the Industrial Revolution and some have survived to this day and are still in use.

Industrial Revolution soon resulted in a substantial increase in demand. The water wheel powered not only the tilt hammers that crushed the ore but also the bellows feeding the furnaces. As a result, the number of ponds in Britain increased significantly. These differed from previous man-made ponds in that they had a flow of water through them and were unlikely to dry up. Many still survive today and teem with life. If they are in soft water areas they have an abundance of dragonfly life and in hard water areas crayfish and bullheads will be found underneath stones together with some species of mayfly nymph. Other mayfly nymphs such as that of the 'greendrake' prefer areas of silt to burrow in. As the water leaves the millpond, there may be a stone channel guiding it. This will be a good place to look for the larvae and pupae of the blackflies and the river limpet.

Canals

Canals have been with us for hundreds of years; indeed some date back to Roman times. Although many are redundant today, they were formerly the main transport highways. Once again the Industrial Revolution saw a vast increase in the length and usage of canals for connecting industry with ports and other areas of demand. In their heyday you could travel the length and breadth of Britain without touching dry land. Compared to transport on the poorly developed road system, canals offered the ability to transport heavier loads in a shorter time. As canals cannot operate over gradients, lock gates were built and there were often many

Figure 15 The Basingstoke Canal in May is a flourishing freshwater environment. Despite clearance and constant use by barges, it is still a very good site for pond life, especially dragonflies.

below each other if the gradient was steep. During the operation of a lock gate a lot of water is lost from the higher reach of the canal, and to replace the necessary volume of water, large areas were excavated adjoining the canal. These 'flashes' also served as turning points for barges. Today they often provide refuges for wildlife especially if the canal is well used.

A canal is, in effect, a long, continuous pond. In general, there is only a slow flow of water and the environment is quite stable. In former years, canals were colonised by animals and plants from the then numerous village and agricultural ponds. Nowadays, as ponds disappear rapidly, canals are becoming increasingly important as wildlife havens.

One problem that ponds and canals have in common is that they eventually fill in. Decaying plant material ultimately forms a stable enough layer for emergent vegetation to grow across the width of the canal. The final stage is for trees such as alder and willow to colonise it which effectively destroys the freshwater habitat and the life it supports. If a canal is to continue to serve barges and provide a suitable habitat, it must be periodically cleaned. This has been done with great success in several areas and is discussed in Chapter 7 on conservation.

In remote areas of Britain, such as Dartmoor, you may come across what look like miniature versions of canals. These are called leats and circle hillsides, often for many kilometres, transporting the water to isolated areas. The Devonport Leat, constructed by Sir Francis Drake, is extremely long, starting at Princetown on Dartmoor and running to Devonport. Leats are often rather barren environments for freshwater life being stonelined. They do, however, support most of the invertebrates found in neighbouring streams. These include the nymphs of stoneflies and caddisflies which in turn feed populations of brown trout.

Figure 16 The Devonport leat on Dartmoor contains the representative plants and animals of the adjacent moorland streams.

Ornamental lakes

The man-made bodies of freshwater discussed so far have all had practical uses. However, wealthy landowners in the past often created ponds and lakes for purely aesthetic reasons. These ornamental lakes were designed to complement their carefully landscaped gardens and were invariably formed by damming a river or enlarging its course. The flow helped them survive and though often deep, certain types of vegetation, such as water lilies, flourished. Fish populations, and in particular the carp family or Cyprinidae, are often large. The bottom-dwelling molluscs form part of their diet and can occur in great abundance. One drawback of ornamental lakes is that it was often thought desirable to introduce wildfowl to them. If ducks and geese are overstocked they can have a harmful effect upon a lake, especially if they are tame, trampling around the edge and creating a high input of organic matter in their faeces. This is particularly evident on some present-day village ponds where mallard or Aylesbury ducks have been introduced. If these are fed, then the pond can support a larger population than it would do under natural conditions. Ornamental ducks are attractive and interesting creatures, but they are not beneficial to aquatic life in a confined area.

Figure 17 Many ponds, such as this one at Scotney Castle in Kent, were created for ornamental reasons to enhance the beauty of an area or property.

Gravel pits

Many freshwater habitats exist as by-products of Man's activities. Some of the most noticeable are gravel pits, which occur predominantly in southern England. Gravel is defined as being approximately 2.5 cm diameter rock particles worn smooth by water. It is widely used in foundations for roads and buildings. In past ages gravel was deposited on river beds and flood plains which is why gravel pits are often found beside the courses of present-day rivers such as the Thames. Here, gravel extraction takes place all the way from Reading to London. Although the layer of gravel may only be thin, it can spread across a wide area. Extraction of gravel causes widespread devastation of the land because the topsoil has to be removed first. Subsequently, however, the pits that have been created generally fill up with water naturally.

Figure 18 A disused gravel pit at Theale near Reading filling in with water. At present it attracts large numbers of wading birds which feed on invertebrates in the small pools.

Gravel pits have the disadvantage of often being deep and steep-sided. Sufficient light for photosynthesis cannot penetrate more than about 10 m and so plants grow mainly around the edges. If the sides are steep then the zone of vegetation may be very narrow. As the pits are often large, the wind may cause significant wave action which will erode the shore as well. Having said this, gravel pits are sometimes purposely landscaped with shelving shores and islands in the middle which increases their value to wildlife significantly.

In these materialistic days everything seems to need a purpose in order to survive. Gravel pits have become popular sites for sports such as yachting, water-skiing and fishing which often cause disturbance to wildlife. Fishermen often stock pits with large numbers of coarse fish and sometimes rainbow trout. Although the prime interest of fishermen in gravel pits is the desire to catch fish, they do appreciate the need for a thriving freshwater environment as without the plants and animals, there could be no fish. They are often the first to notice any change in the water due to pollution.

Other excavations

The gravel pits that we see in southern England are fairly recent creations, but some other large water bodies have earlier origins. During the Middle Ages peat digging was essential for fuel. In north-east Norfolk each parish had peat workings to supply its own needs. In subsequent years the peat workings were flooded and became the Broads, each broad representing the past workings of a single parish. Today the Norfolk Broads are popular as recreational areas but they are also tremendously important wetland habitats. The water contains a wealth of life including a few national rarities, such as the Norfolk aeshna dragonfly found nowhere else in Britain. The channels are lined with extensive reedbeds which form major strongholds for reed-loving birds such as marsh harriers, bitterns and bearded tits.

Coal of course is a more efficient source of fuel than peat. In the Stodmarsh area of Kent former coal workings have subsided to produce an extensive and valuable area of flooded wetland now a National Nature Reserve. It supports similar life to that of the Norfolk Broads but benefits by having much less disturbance from boating.

Figure 19 Reeds often grow over vast areas such as here at Stodmarsh in Kent. This important area for wildlife is a national nature reserve.

In many parts of southern England, especially on the outskirts of sandy agricultural land, deposits of chalky clay or marl can be found. The marl was excavated in the past, and spread on the land to bind and enrich sandy soil and improve its fertility. The series of marl pits that resulted are common along the south coast, particularly in the New Forest area. As these pits are often more than 100 years old, they have been extensively colonised, water bugs, beetles and dragonflies being particularly well represented. In some New Forest pits you can find large numbers of the great diving beetle and bugs such as the water scorpion. In 1979 one particular series of pits held the nymphs and adults of 11 species of dragonfly, including the black sympetrum and the downy emerald.

Since clay contains a good deal of chalk, the land surrounding the pools often supports a wealth of surprising chalk flora. I have seen dozens of autumn lady's-tresses orchids, together with thyme, yellow wort and centaury growing less than a metre from a pool containing all the typical pondweeds. If you have good eyesight, look out for tiny hopping insects. These may well be groundhoppers – relatives of grasshoppers – and they thrive in marl pits. They are relevant to freshwater study since, to escape danger, they can dive underwater and swim for some distance.

Figure 20 In former years marl pits were dug to excavate clay. Many remain today, such as this one in the New Forest, and often contain a wealth of pond life.

In the last 100 years, the population of Britain has expanded tremendously. This has led to increased water demands on the one hand and greater land usage on the other. Large reservoirs have been created to store water for human requirements, but increased land usage has encouraged the removal of water from other areas by drainage. Both practices have led to the creation of new freshwater habitats, as well as the destruction of existing ones. The newly created habitats are, however, not always suitable for our endemic species.

Reservoirs are often found in upland areas since any land used here has less economic value than lowland pastures. Also the rainfall is generally higher and the rock is more likely to be impervious. They may be created either by raising the level of existing lakes or by flooding valleys after damming rivers. In lowland areas, reservoirs tend to be entirely man-made with artificial embankments. In the Staines area of London, there are many reservoirs which are comparable in size with upland lakes.

Upland reservoirs, like lakes, tend to be nutrient deficient or oligotrophic. The life that is present is most noticeable and abundant around the edges and in the surface layers, because of the depth. The diversity of animal life is limited compared to that of a lowland pond, but there may be an abundance of those species present. Lowland reservoirs are similarly oligotrophic and their edges support virtually no plant life except algae. However some have productive planktonic life and are in some ways similar to upland lakes. There is a crustacean of strange appearance called *Leptodora kindtii* which is found in upland lakes and also in reservoirs in the London area. Another similarity is in the importance of these

Figure 21 The waters of this bleak oligotrophic upland reservoir at Craig Goch in Wales lack nutrients and so support little freshwater life.

water bodies to wintering wildfowl, although the lowland reservoirs support greater numbers. Diving species such as the tufted duck particularly favour reservoirs because of the high numbers of molluscs like the zebra mussel that are found there.

Many areas of land that were formerly unsuitable for agriculture have been drained. This is normally done by digging a criss-cross series of ditches which interconnect and draw water out of the land. The ditches provide an ideal environment for many freshwater animals having a slow flow of water which is generally nutrient-rich. The Lewes brooks area of Sussex for example has been extensively drained and there is also a gradation in the salinity of the dykes. The River Ouse into which they drain is tidal for some way and this increases the salinity in the dykes nearest to it. In the most saline ditches in the Lewes brooks, you can find marine algae such as gutweed and estuarine prawns together with the freshwater louse.

Figure 22 A drainage ditch on the Lewes Brooks in Sussex. The water contains an abundance of plant and animal life including marsh frogs.

Further away from the brackish part of the river, you will find purely freshwater plants and animals. One of the delights of this area is the frogs. The common frog is rare but the introduced marsh frog is common, producing its amazing croak during the warm summer months.

3 Getting started: equipment and techniques

Precautions

Before you even begin your pond watching do obtain the permission of the landowner on whose land the pond lies. This may seem obvious advice, but this simple courtesy will not hurt, especially if you want to study the pond on more than one occasion.

Many suitable areas lie on farmland. It may be a pond in the corner of some fields, or drainage channels criss-crossing the land. In most cases, the farmer will be only too happy to allow you onto his land, providing you obey the country code, shut gates and refrain from disturbing livestock. Drainage channels can be particularly rewarding areas of study if they are periodically dredged as you will be able to see the process of re-colonisation at first hand. Indeed you may be able to provide the farmer with information about what is in the water, a rare or local species of dragonfly for example. If you can arouse the interest of the landowner, so much the better; an enthusiastic account of what is present could even prevent his losing interest in a pond or dyke and filling it in. Habitat loss is all too often the result of landowners simply not knowing what is there, or how valuable the site is, rather than wilful destruction.

Your interest in freshwater will probably not be so well received in areas of chalk streams which are frequently owned and managed for trout fishing. Since people may pay hundreds of pounds for a single day's fishing, they do not take kindly to intruders dipping nets into the water. They are often skilfully maintained to support the maximum number of fish and are largely pollution free. As such they are ideal places to go and observe freshwater life, but from a distance. If you are particularly interested in the life in chalk waters, you may be lucky enough to have watercress beds nearby. The life that these support is often similar to that of chalk streams. Again, do not go trespassing in used watercress beds, but ask permission, or better still, try to find a disused one.

Many gravel pits are also stocked with coarse fish or trout and belong to angling clubs. Although many owners will refuse you access, a surprising number may allow you to visit a quiet corner. Other gravel pits may be used for sailing or water-skiing. Again,

Figure 23 Watercress being gathered in a Hampshire watercress bed. Disused beds can often be rewarding places for pond study.

the same thing applies about permission; but the owners are generally less possessive about their water.

Most village ponds, canals and heathland areas have public access and you will be free to visit them. However, do be careful not to interfere with the interests of other people. Both canals and heaths present a very real danger of drowning. Canals can be treacherous because of their steep sides, and this is often the case with gravel pits. Heaths are potentially dangerous because of the boggy nature of the ground. Their soil and vegetation is very porous and soaks up

Figure 24 Bogs can be dangerous places to explore because *Sphagnum* moss often grows over open water. Not all bogs are as clearly marked as this one, so do take care.

water. Sometimes the vegetation grows over the surface of free water creating a floating bog. One of the typical plants colonising floating bogs is *Sphagnum* moss so be very wary of treading on it. In this habitat it assumes a fresher shade of green. If in doubt about approaching a heathland pool, then the rule is don't. You may end up immersed in water up to your waist or further! If you are unfamiliar with the habitat then go with someone else and take especial care.

The more usual hazard of pond dipping is getting soaking wet or plastered with mud. If you use a certain amount of commonsense and wear suitable clothing, this can be avoided. Wellington boots or waders are the obvious footwear. If you do not mind getting wet you can simply roll up your trousers. Do wear plimsolls though because there may be broken glass buried in the mud. Glass and barbed wire are the two major problems for wellington wearers; I have lost more pairs because of them than anything else.

A last cautionary note is about avoiding causing damage to the environment. Again, this is a matter of commonsense. The banks of rivers and ponds can often be damaged by foot erosion. If they are steep then it is all too easy to cause a landslide. Where the banks are not as steep, try to trample as little vegetation as possible. The plants not only provide cover for certain animals, but also stabilise the bank and prevent natural erosion.

If you dredge a sample of weed onto the bank, then please return it when you have finished. Aquatic plants are generally unable to survive out of water because they cannot cope with desiccation. The plants also provide a home for numerous animals, some of which will be trapped in the weed. In streams and rivers stones often do the same so if you are working there and turning stones over to look for animals, please turn them back when you leave.

Equipment

The most familiar item of equipment used by the freshwater naturalist is the net. Small nets can be bought from an aquarist shop and these are useful for catching slow-moving animals which are close to the pond edge. They are obviously impracticable for deeper waters and for collecting large quantities of plant material as they bend and break very easily.

A larger net will help here. Again, these can be bought, but it is also quite easy to make one yourself. You need a stout pole which will stand up to the rigours of pond dipping such as an old broom handle about 1½ m in length. Draw a line across the diameter of the flat end of the pole with a pen and continue the lines down both sides of the wood. These should then be diametrically opposite each other. Nail in three chicken wire staples down each line at 5 cm intervals. These will eventually secure the wire frame of the net and you should choose staples large enough to hold the wire in place. The total length of the wire should be about 90 to 100 cm and its diameter about 8–10 mm. Bend the wire to form a loop with the ends parallel; imagine that you are outlining the shape of the head and neck of a tennis racket. The two ends of the wire loop should then fit through the staple loops on either side of the handle. Before you do this, however, the net must be made and fitted.

The mesh size of the net should be about 1 mm. If it is larger animals will escape and if it is too small, it will clog up with silt in no time and not allow water to pass through. Nylon is a good material to use. First of all cut out a triangle of net with a base of about 90 cm and a height of about 50 cm. Fold this in half along the base and then cut off the top 10 cm of the triangle. (This will eventually form the end of the net.) If you do not cut the end off, the net will be too narrow and will trap animals and silt, making the contents difficult to examine. Next, sew up the side of the triangle with several layers of stitching, preferably using a sewing machine. It is a good idea to strengthen the join by sewing a strip of protective cloth around the edge. Nets snag easily when dragged through water and can rip your sewing. Remember to leave the 15 cm nearest the base of the triangle free as this will be used to fit the net around the wire frame.

The next stage is to form the loop which will thread around the wire frame. Fold back the edge of the base of the triangle and sew it down allowing a wide enough gap for the wire to pass through easily. Again, you can strengthen this by sewing on another layer of material remembering to leave the ends open.

Thread the net round the wire frame and down its two sides. Finally, insert the ends of the wire through the staple loops and hammer these home to grip the wire. The net is now ready to use but one further piece of protection can be added. Cut a length of rubber tube equal to the net's circumference (about 90 cm). Slit this down one side along its whole length and wrap it around the net and wire framework to prevent the rim of the net from being damaged or torn. The tubing can be fixed in place by sewing through the net with stout thread or fishing line at 3 cm intervals and tying the thread tightly to secure it. (See Figure 28 on page 38 for diagrams.)

If you are lucky enough to have a microscope and are interested in planktonic life, then a special net can be made to sample the water. The frame and the initial triangle of the net can be the same as just described but the mesh should be much smaller. Do not cut off the top of the triangle, and when sewing it up leave a small opening there. Sew a strip of material with long free-ends around this

---

Figure 25 Pond skaters live on the pond surface relying on surface tension to keep them afloat. They feed on animals trapped in the water.

Figure 26 Water fleas or *Daphnia* are often abundant in the summer months in ponds and provide a food source for many of the larger organisms.

opening. These can be tied around the mouth of a small bottle placed through the opening to secure it. The microscopic animals will be unable to pass through the fine mesh and will collect in the bottle. They are thus unharmed and the bottle can easily be removed and its contents examined.

Another useful piece of equipment, particularly for collecting weed samples, is an old rake. If it has a long handle, you can simply drag weed in from the bank. Alternatively, if you remove the metal rake and tie it to a length of rope it can be hurled into the pond and dragged in again. Plants from deeper areas can be collected in this way, but be careful not to cause too much damage.

If your net samples are muddy, or if you want to study the life within the bottom sediments, a sieve is invaluable. Several species of dragonfly nymph and numerous worms and molluscs live exclusively in this habitat. They can be very difficult to spot unless they are 'cleaned-up'. An old kitchen sieve is perfectly adequate but it does not hold a large quantity of mud. I have constructed larger sieves which have proved more useful. The ideal container is a large ice-cream or margarine tub. Cut out the bottom, leaving a margin around the outside of about 1 cm. Then cut a piece of metal gauze to fit tightly inside the tub. This will then rest at the bottom, supported by the lip that you have left. The mesh size of the gauze depends on the size of the organisms that you are looking for. However, the coarser the mesh, the easier it will be to use the sieve.

A sample of mud will never run straight through the sieve, and you will have to work quite hard to clean the animals. The most effective way of doing this is to immerse the tub in water almost to the brim. Water will rush in through the gauze and percolate through the mud. If the tub is then lifted above the surface, water will pour out, taking some of the mud with it. By repeating this procedure several times you will end up with a few stones and debris such as caddis cases and snail shells. But, more importantly, amongst these will be your mud-dwelling organisms. Another advantage of sieving this way is that any animals that pass through the sieve return immediately to the water.

On reaching your chosen pond, before you even begin to dip, consider where to put your samples once you have caught them. Pond organisms are, by their very nature, dependent upon water. If you remove them from this environment for any length of time some of them will inevitably suffer or die. Initially the best thing to put them in is a white enamel tray. Large margarine tubs are ideal substitutes and, being plastic, are lighter to carry. Your sample can be sorted out and the specimens that you want to study further

Figure 27 Crayfish are found in chalk streams and spend much of their time under stones. This one has adopted an aggressive posture and is brandishing its large pincers.

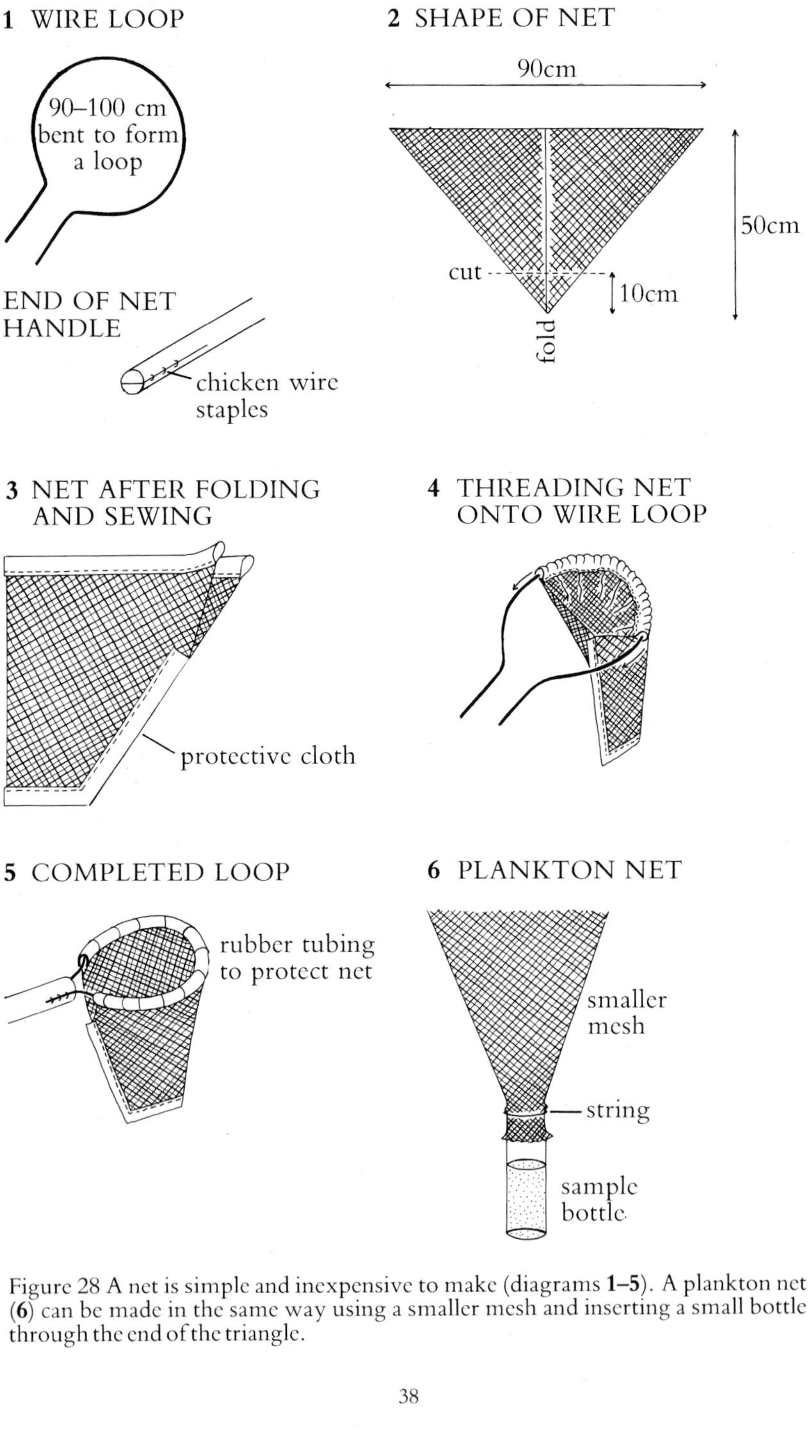

1 WIRE LOOP

90–100 cm
bent to form
a loop

END OF NET
HANDLE

chicken wire
staples

2 SHAPE OF NET

90cm

50cm

cut

10cm

fold

3 NET AFTER FOLDING
AND SEWING

protective cloth

4 THREADING NET
ONTO WIRE LOOP

5 COMPLETED LOOP

rubber tubing
to protect net

6 PLANKTON NET

smaller
mesh

string

sample
bottle

Figure 28 A net is simple and inexpensive to make (diagrams **1–5**). A plankton net (**6**) can be made in the same way using a smaller mesh and inserting a small bottle through the end of the triangle.

Figure 29 Your pond dipping equipment should include a long handled net, an assortment of plastic buckets, a sieve, some smaller nets, an enamel tray, a spoon and a pipette. A plastic sheet will also come in handy for sorting out samples of weed.

transferred to smaller tubs or plastic bottles. Try to avoid using glass bottles since they are easily broken, especially when your hands are wet and cold. Broken glass can not only cause injury to yourself, but also to other unsuspecting pond hunters at a later date.

It is a good idea to carry a plastic sheet with you to lay on the bank with one edge in the water. Large samples of weed can be placed on this and the animals quickly sorted out. After you have finished, the whole lot can be rinsed in the water. This means that no animals will be stranded on the bank and left to dry up.

These are the major items of equipment needed for successful pond dipping. However, there are a few useful accessories. A supply of plastic spoons is a must as it will enable you to transfer soft-bodied animals from one container to another without damaging them and means that you can avoid picking up biting animals such as water boatmen. For smaller animals a pipette such as an eye-dropper is useful. Small crustaceans such as ostracods and copepods are really difficult to handle alive without one. Lastly, a hand lens opens up a whole new world. You will be able to see such intricacies as the gills of mayfly nymph or the mask of a dragonfly nymph.

Getting wet is an inevitable consequence of pond dipping. If you have any equipment with you that you want to keep dry, such as a camera or hand lens, then it is useful to carry a towel with you. Although this is seldom done, it can be invaluable. (Photography is discussed fully in Chapter 6.)

With all your trays and bottles laid out on the bank, you can now approach the problem of getting your samples. There is a certain art to using a net of any description. When sampling in open still water,

a

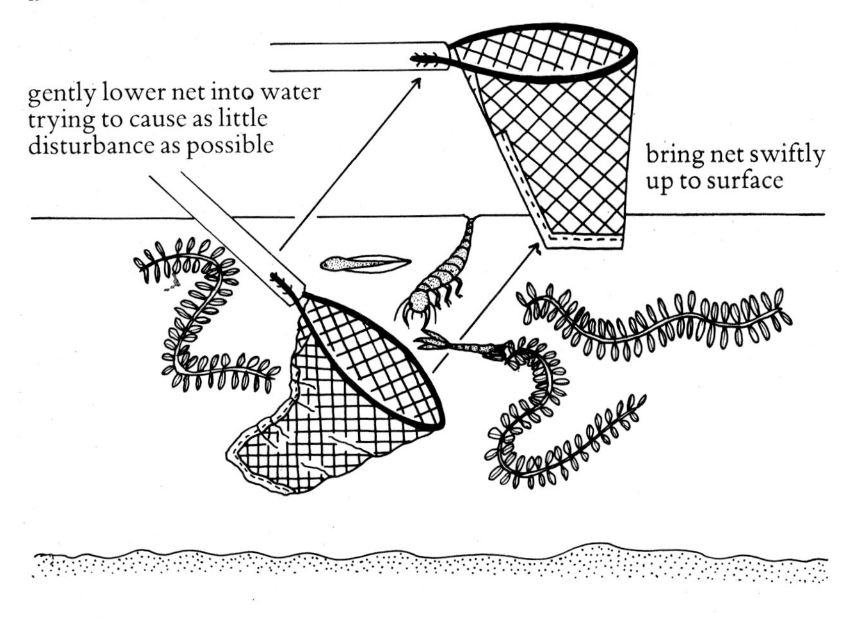

b

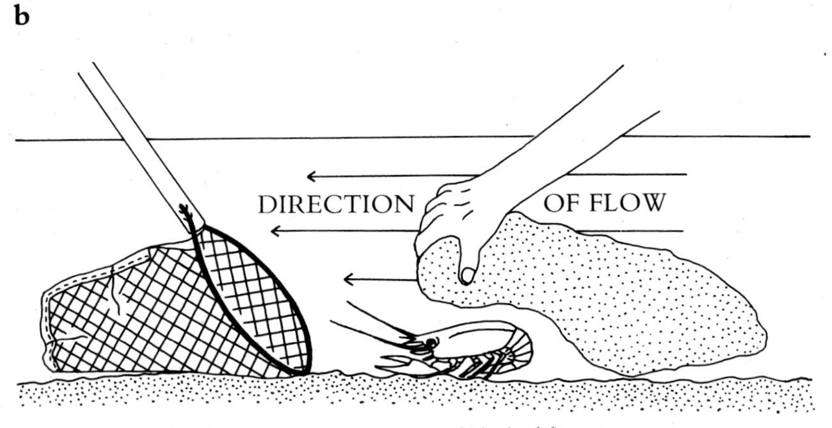

lift boulders and stones while holding net
downstream; animals will be carried into net

Figure 30 The techniques for using a net vary according to the strength of the current. When using it in still water, lower the net gently trying to cause as little disturbance as possible and then bring it swiftly to the surface (**a**); in flowing water hold the net downstream and as you lift up boulders and stones, animals will be carried into the net (**b**).

a figure of eight pattern often produces good results. Do remember to maintain an even sweeping speed. If you suddenly stop in mid-water most of your catch will escape. When sampling amongst weeds it is a good idea to lower the net to the bottom and bring it to the surface swiftly. This gives less opportunity for active swimmers to escape. If you want to catch dragonfly nymphs, try sweeping the same area several times. Disturbed from the surrounding weed, they will attempt to escape and may swim into the area that you have just netted. Your disturbance will have removed some weed and swept more aside, leaving a temporary patch of open water. You therefore stand a better chance of catching a swimming nymph here. Never be tempted to collect too much in your net in one go. This may damage it and in any case, it will give you too many animals to sort out in a short time. If you are netting amongst weed, an excessive quantity will probably be too entangled with its surroundings to be lifted out and you will end up having to tip the whole lot back!

Using the net in running water requires a different technique as you must consider the direction of the current. If you are netting amongst vegetation, work the net upstream so that any dislodged animals will drift into the net. Many stream dwellers can be found under stones. A good way to catch the more active ones is to hold the net downstream of the stone. Then lift the downstream end of the stone and any animals trying to escape will naturally get swept into the net. Please remember to replace the stone after you have finished.

Approaches to pond hunting

You should now have all the equipment needed for a successful day's pond study. However, the way in which you approach this study can make the world of difference to its outcome. Your two best attributes will be a tolerance of cold water and a sharp pair of eyes. Do not forget that the inhabitants of the pond will be on the lookout for danger. Many pond animals detect predators and other dangers by vibrations in the water so a stealthy approach to the pond's edge is essential. Try not to move too quickly once in the water, because vigorous movement either of your wellingtons or your net will only drive away the more active animals.

Shadows cast on the water are another indication to pond dwellers that danger is near. If possible, approach the pond with the sun in your eyes keeping as low as you can to avoid the problem of shadow. If the edge of the pond has a steep drop, then try lying down and fishing through the water – a much more relaxing way to go about it.

A certain amount of patience is required when pond hunting if you want to get the best results. Many animals are very good at

Figure 31 This group of pond dippers has arranged its equipment on a plastic sheet so that the remains can run back into the water from samples. While two people net from the water, a third sorts through the plant samples in an enamel tray.

concealing themselves if disturbed. This is particularly true of *Hydra* which resembles a small sea anemone. When disturbed it contracts its body into a blob. If you leave a sample of pondweed in a jar for a few minutes, many *Hydra* and other animals will appear, and you will be amazed at what you missed on your first look.

There are certain times, both of the day and the year, when a visit to a pond is most rewarding for specific creatures. During the morning the temperature rises and insects such as dragonflies become more active above the water. Indeed, on a cold day they may not take to the wing at all. In the water itself, the metabolism of the inhabitants increases with temperature as does their oxygen requirement. Those creatures that breathe aerial oxygen such as water beetles will return to the surface more frequently in warmer water. Thus the middle of the day is a good time for pond hunting. Early mornings are, however, better for some other animals. Bird life is typically more active at dawn than during the day. Also an early morning visit may reveal the wonder of dragonfly emergence. This is best seen in late spring when a search of the vegetation surrounding the pond is sure to reveal a newly emerged insect or its cast nymphal skin. Spring is also the time when you must exercise some caution. It is the breeding season for birds and the spawning time for fish and amphibians. Try not to disturb areas where this is taking place. It would be a shame if, whilst pursuing your pastime, you destroyed something that contributes to the life of a pond.

The other major *don't* is not to disturb the pond when there is ice on the surface. As explained earlier, the ice acts as an insulator against heat loss. The animals underneath never normally experience freezing conditions and might die if the ice were broken. I hope that these words of advice and caution will enable you to be more productive in your pond hunting. You should then be able to pursue and enjoy this fascinating hobby, causing the minimum of damage to the environment.

The aquarium

One of the most absorbing and relaxing ways of studying pond life is to have a freshwater aquarium at home. This need not be elaborate or stocked with unusual animals and plants, for you will find that even the most humble of pond dwellers has an interest all of its own. There is something very satisfying about an aquarium that you have nurtured into a stable environment.

The tank can be bought, or better still, homemade to your own specifications. A good size is 30 x 30 x 60 cm which allows a good surface area in relation to the volume of water in the tank. The oxygen requirements of the organisms can be satisfied by diffusion from the air. The surface area to volume ratio is important and is one reason why goldfish bowls should be avoided as the air simply cannot supply enough oxygen if it is well stocked with animals.

Constructing your own tank is a fairly simple affair with the use of silicone rubber aquarium sealant which will support a full tank of water of the dimensions mentioned, and is perfectly watertight. Plate glass 0.5 cm thick is the best material to use. The base and the two longer sides should be cut to exactly 30 x 60 cm. The two shorter sides should be 30 x 29 cm – this will allow the longer sides to fit neatly on the base outside them.

Lie the bottom sheet of glass down; all the joints will be on its upper surface. The aquarium sealant usually comes in tubes and an even layer should be applied along the upper surface of the shorter edges and two smaller sheets of glass placed on the top with the 29 cm side on the sealant. Ensure that each side rests on the bottom glass with a 0.5 cm gap at either end. The two ends can now be supported by heavy weights on either side. Next, apply the sealant along the remaining edges of the base and ends of the tank. Place the two longer sides in position pressing them firmly against the base and the sides. Be generous with the sealant since water will leak through any gaps. There should be sufficient to ooze out of the joins and any excess can be smeared across the joins to ensure a watertight fit. Tape up the joins to support the glass while the sealant hardens. This takes about a day, after which the tank is almost ready for use. At this point you can remove any excess sealant with a razor blade.

Put the tank outside and fill it with water. This will allow you to check for leaks and will also remove any acetic acid which has been released by the sealant during the hardening process. If any leaks

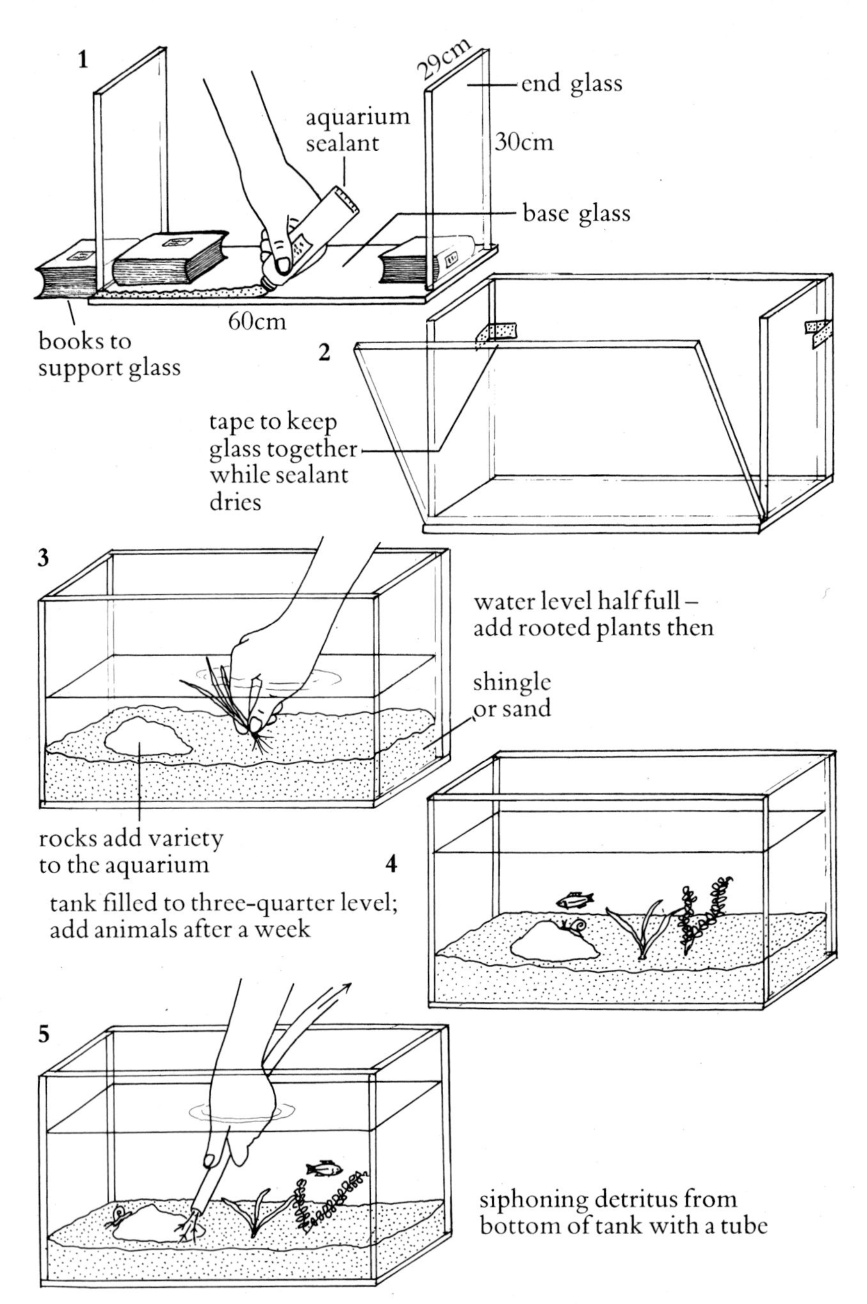

Figure 32 By making your own aquarium at home you will be able to study freshwater life at close hand and many suitable projects are suggested in Chapter 6. Do remember to clean the tank regularly by siphoning away the detritus (**5**).

should appear, drain the tank, dry it and apply sealant to the relevant spot.

If you are constructing a tank for photographic purposes only, then it is best to make it considerably smaller. It is also a good idea to have one side made from high optical quality glass so that the photographic image is undistorted. Tank photography is dealt with in greater detail in Chapter 6.

Before you fill and stock the tank, you should decide on its final resting position. It will be extremely heavy and difficult to move afterwards and you may disturb some of the animals. A sheet of foam rubber underneath helps to protect the surface on which it is standing from the sharp edges of the glass. The tank should be near a window but must never be placed in direct sunlight. This would heat up the water, which in turn would increase the amount of oxygen required by the tank's inhabitants. Sunlight would encourage algal growth, turning the water into pea-green soup. A north-facing windowsill is ideal and preferably the room it is in should not be centrally heated. Many of the larval stages of freshwater animals use temperature as one of the indicators of season. If they are kept continually warm, some may hatch out in the dead of winter. These individuals will inevitably die if released outside and in any case would never find a mate.

Having decided upon the tank's final position, a 2.5 cm layer of fine gravel should be placed in the bottom. Sand can be used but it is more difficult to clean. This is an important consideration because dead plant and animal remains and faeces collect on the bottom very quickly, gradually working their way in amongst the substrate. So the larger the stone particles, the easier the periodic cleaning becomes. If the gravel has been bought from an aquarist, then it can be used immediately. If you have collected it from the beach, it must first be soaked for a day to remove the salt and then washed to remove any organic debris. This is best done in a bucket where the gravel can be swirled with a stout stick and after it has settled the water can be poured off. This should be repeated several times to ensure perfectly clean gravel. A few large and jagged stones will improve the tank's appearance. Not only will they give it added visual interest by providing a three-dimensional environment for crawling animals, but will also provide shelter and cover for others.

The next stage is to start filling the tank. Initially this should be to a depth no greater than 8 cm. You can use rainwater or tapwater but the latter should be allowed to stand for a day for the chlorine, used in purification, to evaporate. Better still, if you are stocking your aquarium from one particular pond, why not use the water from this source? The animals and plants will be much happier in it. If you do not want to disturb the gravel when you pour in the water, then place a plastic sheet over it and pour the water onto this.

When the water level has reached the 8 cm mark, it is time to add the rooted plants. Most of them are buoyant and must be anchored down. Try to find a suitable jagged stone and tie thread around this and the base of the plant. Then bury the stone and roots in the gravel. The selection of plants and animals for your tank is particularly important. Before you make your selection, read Chapter 5 on freshwater ecology. Not all plants and animals survive in the same types of water. For instance, it would be no good trying to keep snails and crustaceans in water from an acid peat bog. So please give careful consideration to the stocking of your tank.

Amongst the rooted plants, the pondweeds *Potamogeton* spp. are particularly good and some species will produce floating as well as submerged leaves. Water lilies are generally too large for the average tank, but frogbit is a smaller version and can be a nice addition. This species dies back in the winter after producing over-wintering buds. You may need to clear it out in the autumn to prevent it from fouling the tank. The buds can be buried in the gravel and will grow the following spring.

With your rooted plants in position you can now fill up the tank. Do not fill it more than three-quarters full as this gap is essential for the gas exchange between water and air. Many floating plants can now be added including Canadian pondweed, hornworts and starworts. These plants do not normally receive the same intensity of sunlight as terrestrial plants, because the water absorbs it. Consequently they do not do as well if the tank is in bright sunlight, another good reason for placing the aquarium on a north facing windowsill. All of the plants just mentioned are readily available from aquarists' shops or from garden centres. As it is illegal to uproot plants without the permission of the landowner, it is a better idea to buy these when stocking your tank.

With the plants established, you can now choose the animals for the tank. You will find yourself torn between those with an attractive appearance or behaviour pattern, and those which would be beneficial to the aquarium environment. A good mixture is the best compromise. Snails are useful since they feed by 'vacuum cleaning' the bottom and grazing algae. However, beware of having too many large snails because not only do they eat parts of living plants, but their excreta also provides a perfect substrate for bacteria to feed on, and the latter may get out of hand. A certain amount of decay is essential and will help recycle nutrients, but too much can turn the water anaerobic. One or two great pond snails and ramshorns are ideal, but watch out because they multiply very rapidly. Small freshwater pea mussels will happily live in the detritus at the bottom and will also help to clean the tank.

Freshwater insects provide a more lively display. Be careful to select those species adapted to standing water, because animals

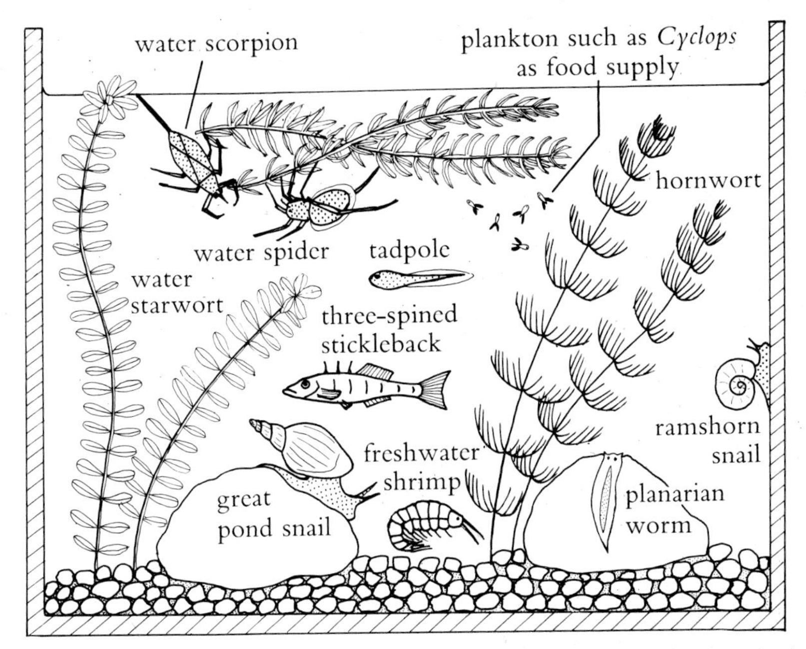

Figure 33 When stocking your aquarium try to combine animals which look attractive with those which have interesting behaviour patterns. Choose a selection of rooted and surface floating plants and buy these from an aquarist. All the plant and animal species shown here would be suitable for your tank.

adapted to flowing water generally have a higher oxygen requirement and will die in the aquarium. Caddis larvae are rather endearing as they trundle around carrying their stick cases. Very few species of mayfly live in standing water, but the pond olive is suitable for your tank.

Many species of water bug and beetle survive happily in an aquarium. The water stick-insect and the water scorpion, although predatory, are fascinating to watch and will take only relatively small prey. Larger animals such as the water boatman will attack quite large prey and you may have to decide whether the disappearance of some of the larger inhabitants of the aquarium is acceptable. One animal that should definitely be avoided is the great diving beetle. Both the larva and the adult are voracious and will eat all the other animals in the tank in no time at all.

Dragonfly nymphs are also predatory. The smaller damselflies take proportionately smaller prey, while the larger dragonflies may tackle prey as large as tadpoles. However, I think that they make up for their diet with the interest that they provide. Larger nymphs such as those of the emperor dragonfly may take several years to mature. They eat rather infrequently and may not make a tremendous impact on the tank's inhabitants. In the spring a few twigs should be placed in the tank so that they rise above the surface of the water. These are essential for those terrestrial insects that emerge from aquatic larvae. Many of them have to climb out of the water to do this.

The life of the water spider can be studied in a tank at home. If left alone in a tank with sufficient weed, it soon builds an air-bell. The tank should, of course, be stocked with a few freshwater invertebrates to act as a food supply. After mating has occurred, the water spider constructs a special tent in which the eggs are placed. In due course the young hatch and live in this air tent for some time. The mother cares for them and guards against predators.

You must also provide a range of smaller organisms whose diets range from detritivore to carnivore. Crustaceans such as ostracods, the freshwater shrimp and the freshwater louse are very suitable. They will help clean up the tank and themselves form part of the diet of more predatory creatures. Leeches and planarian worms are both useful additions to the tank. They help clean up dead remains and are active, providing visual interest.

Smaller creatures such as *Cyclops* spp. and water fleas are also good animals for an aquarium because they form the first level of the food web. Their numbers are often depleted rapidly in an aquarium so it is worth culturing them separately and adding portions to the main tank. Water fleas will feed and multiply rapidly on a diet of planktonic algae such as *Chlorella*. The two can be cultured together by placing a few individuals of each in a tank of water containing liquid 'plant food'. By enriching the nutrients in the water, the algae multiply, turning the water into a pea-green soup, enabling the water fleas to increase. Another advantage of culturing these separately is that you can obtain a large number of water fleas in this way without clouding your aquarium.

It is a good idea to clean the tank every two weeks. This is not a lengthy process and simply involves removing some of the dead plant and animal remains that have collected on the bottom. Some detritus is essential to the successful survival of a freshwater tank, but an excess can cause problems with bacteria build up. Immerse a 1-m length of tubing in the water and allow all the air to escape. Then place a finger over one end and remove it from the water. When this end is lowered below the level of water in the tank, and your finger is removed, water will siphon out. Put the lower end in a bucket and move the other end over the surface of the gravel to suck up the detritus. Since many animals will be living amongst this detritus, you can sieve the water as it comes out and return the animals to the tank. Top up the water level with rainwater. This is essentially free of dissolved salts and will prevent the concentration building up in your tank. For this reason, rainwater should also be used when evaporation from the tank has occurred. Evaporation can be prevented to a certain degree by having a glass lid for the aquarium. This will also prevent bugs and beetles from escaping, which they frequently do.

Fish and newts are often popular amongst tank enthusiasts. However, in a tank as described, I would not recommend any animal larger than a stickleback. One or two will survive in a tank of this size together with some newts. If the aquarium is a miniature version of a pond you will have to accept that some of them will die or be eaten. This occurs all the time in the natural environment and is a normal part of pond ecology. If you want to enjoy watching the courtship and egg-laying behaviour of sticklebacks and newts without them being eaten, it is best to have a separate tank. Their behaviour and study is described in Chapter 6 on pond projects. I once kept an eel in a tank, and far from being dull, it was a most rewarding pet. It came to trust humans and would leave the cover of its pipe to come to the surface to take food from a pair of forceps. It had a most tenacious grip and would never let go of the food.

I have not mentioned frogs in this section because they require a specialised tank where they can spend time out of water. If you want to study frogs, why not rear tadpoles (see Chapter 6) and then release the froglets when they have developed?

Having mastered the techniques of studying pond life, you can now go into the field. The next chapter will help with field identification. When you are familiar with the animals and plants in your pond, you can then carry out some of the detailed projects described in Chapter 6.

4 What to look for

The wealth of animal and plant life in freshwater, and particularly its diversity, is astounding. As such it would be impossible to do justice to the identification of freshwater life, even that found in British ponds, in this book. There are many field guides currently available which will help you (see Further reading). In this chapter I have tried to select and describe in note form the commonest and most distinctive animal and plant groups which illustrate all their key features. This should form a basis for your interest in the practical art of watching and understanding a pond.

Plants

Plant species in ponds are zoned from truly aquatic plants in water to wet-loving but terrestrial plants around perimeter. Ponds are doomed to fill in eventually; by looking at zonation a condensed version of this succession from pond to dry habitat will be clearly evident.

Pond surroundings

Pond eventually becomes fixed stable woodland. This climax vegetation in heathland areas is **birch** scrub with layer of plants beneath such as **bog myrtle**. **Alder** and **sallow carr** common elsewhere. Trees growing in wet places may find difficulty obtaining sufficient nutrients which may be lost in water. To overcome this many have evolved associations with other organisms (symbioses). Alder has root nodules containing nitrogen-fixing bacteria. Birch uses fungal association, the extensive system of branching hairs of fungus (mycorrhiza) increasing surface area of birch roots to help take up of nutrients. Alder and sallow in particular are important food plants for many insects and are vital members of freshwater ecosystem.

Pond margin

There is an increase of lush vegetation from the drier areas to pond margin. Typical plants: **great willow herb** (codlins-and-cream) with deep pink-and-white flowers which contrast with creamy flowers of **meadowsweet** and vivid purple of **purple loosestrife.** **Marsh marigold** also common; a member of buttercup family,

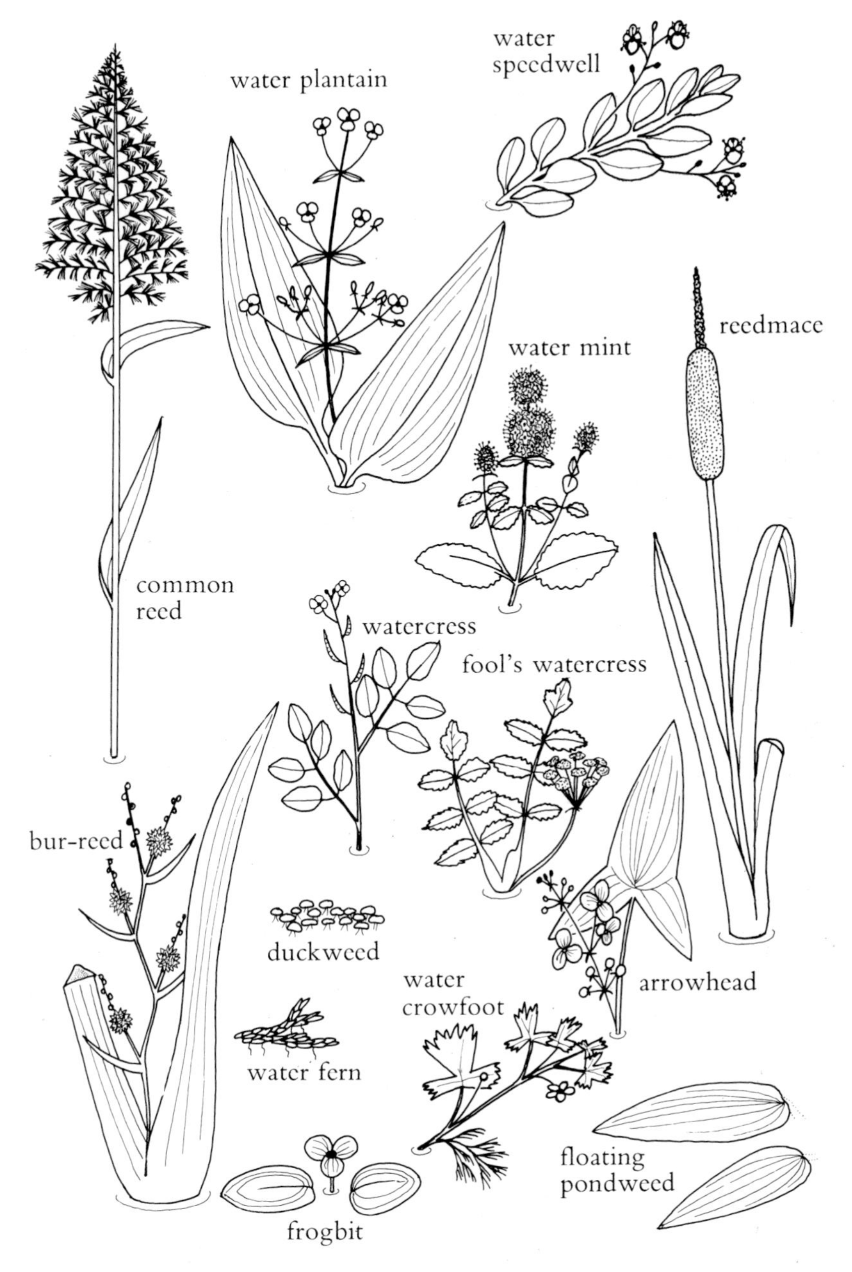

Figure 34 Some of the commonest floating and emergent water plants.

with flowers like larger versions of this plant. Primitive plant with no petals and large, yellow sepals. (Sometimes known as kingcup.) *Juncus* spp. (rushes) may grow in dense stands. Many species in Britain, often difficult to distinguish apart. Phrase 'rushes are round and sedges have edges' is helpful referring to plant stem.

Acidic, heathland areas have pond margin of *Sphagnum* moss and occasional **sundew** plant. Sundews gain extra nitrogen by being insectivorous. A coating of sticky hairs on leaves attracts insects by glistening appearance and traps them. Leaves also release digestive enzymes which break down body of insects; any nutrients are later absorbed by leaf.

Emergent plants

These grow in water around edge of pond and are the flowers conjured up when you imagine a pond in summer. Stunning flowers of **yellow flag** are familiar. This and **sweet flag** extensively introduced: former for flowers and latter for sweet-smelling leaves. Used in medieval times in houses and churches to spread on the floor and provide a more wholesome aroma.

Confusing array of grass-like plants. **Great reedmace** *Typha latifolia* has brown, sausage-shaped flower head. Often incorrectly called **bulrush**, which is in fact sedge *Scirpus lacustris*. **Branched bur-reed** somewhat similar; leaves resemble those of irises but 3-sided and not flattened near base.

Reed *Phragmites communis* If pond is big enough there will be areas of reed. Stands often quite extensive and may even be encouraged where reeds used for thatching. This is the major plant responsible for silting-up of freshwater habitats, as its wide-branching root system traps particles of silt. Stems die back each winter and new growth builds on remains the following year creating a higher and higher reed bed.

Numerous insects feed on it such as larvae of wainscot moths. Provides shelter for nesting birds such as bitterns and marsh harriers; small passerines, such as reed warblers and bearded tits, nest in it.

Reed sweet-grass Similar but does not grow as high and forms less extensive stands. A favourite food of cattle. Growing amongst it will be **horsetails**, extremely primitive, non-flowering plants with many species and many different forms of the same species (ecotypes), which vary according to exact position in which they happen to be growing. Horsetails grew to incredible sizes in prehistoric times and their breakdown and compaction produced

---

Figure 35 This female southern aeshna dragonfly is ovipositing in a rotten tree root at the edge of a pond.

Figure 36 The wings of the demoiselle agrion damselfly have a brilliant metallic sheen. Groups of males can often be seen flying up and down slow moving streams.

coal measures and oil.

Further into water there will be **water speedwells, brooklime** and **water mints**. Many closely related species and hybrids of water speedwells and water mints but two genera are quite distinct. Water mints have a scent and four pinkish petals; water speedwells have four unequal petals and are unscented.

Watercress Often abundant on wet areas on chalk, particularly drainage ditches. It tastes delicious and is rich in iron and Vitamin C. Beware of collecting it if sheep or cattle are nearby, as larval stage of liver fluke encysts on outside of water plants such as watercress to increase its chances of being eaten. If eaten by humans, it develops nicely in the bile-duct! Exercise caution and if in doubt blanch before eating it or turn into soup.

Fool's watercress or **water parsnip** Closely resembles watercress. A member of carrot or umbellifer family whereas watercress belongs to cabbage or crucifer family. Called fool's watercress as it is poisonous and you would be a fool to eat it. **Water dropworts** are also poisonous umbellifers, notably **fine-leaved** and **hemlock water dropwort**.

Bogbean, water plantain and **amphibious bistort** All are found in emergent zone and have somewhat similar leaves. Bogbean is so called because its leaves resemble those of broad bean. Characteristically in groups of three, these are held together above water surface with attractive pink-and-white flowers. Bogbean seldom rises more than 15 cm above water; water plantain may be 60 cm tall. Although unrelated to land plantains, its leaves are broad and similar in appearance. Amphibious bistort is one representative of a number of similar species. It is a more trailing plant than dock or bogbean with pink flowers in a tight head.

Surface-floating plants

Floating plants occupy the important interface between air and water, reaping to a certain degree, the benefit of both worlds. They gain more oxygen from air and more light by remaining on the surface, yet are also kept humid and supplied with nutrients by the water.

Duckweed A typical surface-floating plant with several species. No true stems or leaves but a frond which may be composed of both. Although a flowering plant, flowers are minute and rarely seen in Britain. Main means of reproduction is vegetative division; often by late summer it forms carpet across pond surface which significantly reduces amount of light penetrating water and hence the photosynthesis of plants below the surface. Carpet also provides

Figure 37 Whilst mating and egg-laying, damselflies fly around in tandem. Here a female large red damselfly is laying eggs underwater while the male clasps her behind the head.

plenty of shade and shelter for surface animals and helps modify extremes of temperature during summer.

Frogbit Also forms a carpet as do many floating plants. Seems to disappear completely during winter months, surviving as overwintering buds in the mud. In spring these produce kidney-shaped leaves which redden as they age and look like miniature lily leaves although they are much smaller. Attractive flower is quite different with three white petals supported by leaves.

Water lilies Familiar to most of us. Plant is well adapted to life in freshwater; stems and leaves have air-filled vessels enabling them to float. Cuticle and stomata on upper surface of leaves enables exchange of gases with air and nutrients with water. Propagate by forming rhizomes.

Pondweeds Many species occupy both surface and water column. **Broad-leaved** or **floating pondweed** is characteristically on surface. Propagate by forming rhizomes.

Water fern Two species in Britain, both introduced from America; now widespread. Have interesting symbiotic relationship with blue-green alga which lives in cavities in plant and may have a role in fixing nitrogen and supplementing fern's supply of this essential nutrient.

Submerged plants

Submerged plants are truly aquatic and would usually not survive out of water.

Pondweeds One of the most typical groups. Large number of species which are hard to tell apart. All have alternate leaves except the closely related **opposite-leaved pondweed**.

Canadian pondweed Introduced species which has spread so dramatically that few suitable water bodies are without it. Even a small fragment can produce a new plant; its propagation and dispersal are often aided by efforts to clear it. A favourite plant for garden ponds as it produces a relatively large amount of oxygen. On sunny days it can be seen glistening with bubbles of oxygen which gradually rise to the water's surface.

Bladderwort An extraordinary plant of heathland pools. Acid waters are often poor in nutrients such as nitrogen and plant supplements its diet with water fleas. Bladders, from which name derived, have sensitive hair attached to capped opening which acts as a trigger. If set off by passing water flea, the cap will be flung open, the inrush of water sucking in victim. Whether as a result of digestion or decay, nutrients from water flea are absorbed by plant leaving only exoskeleton.

Water crowfoots These survive equally well underwater and on pond margin. Leaves finely divided however, and quite different from those on same plant growing on surface. Flowers produced

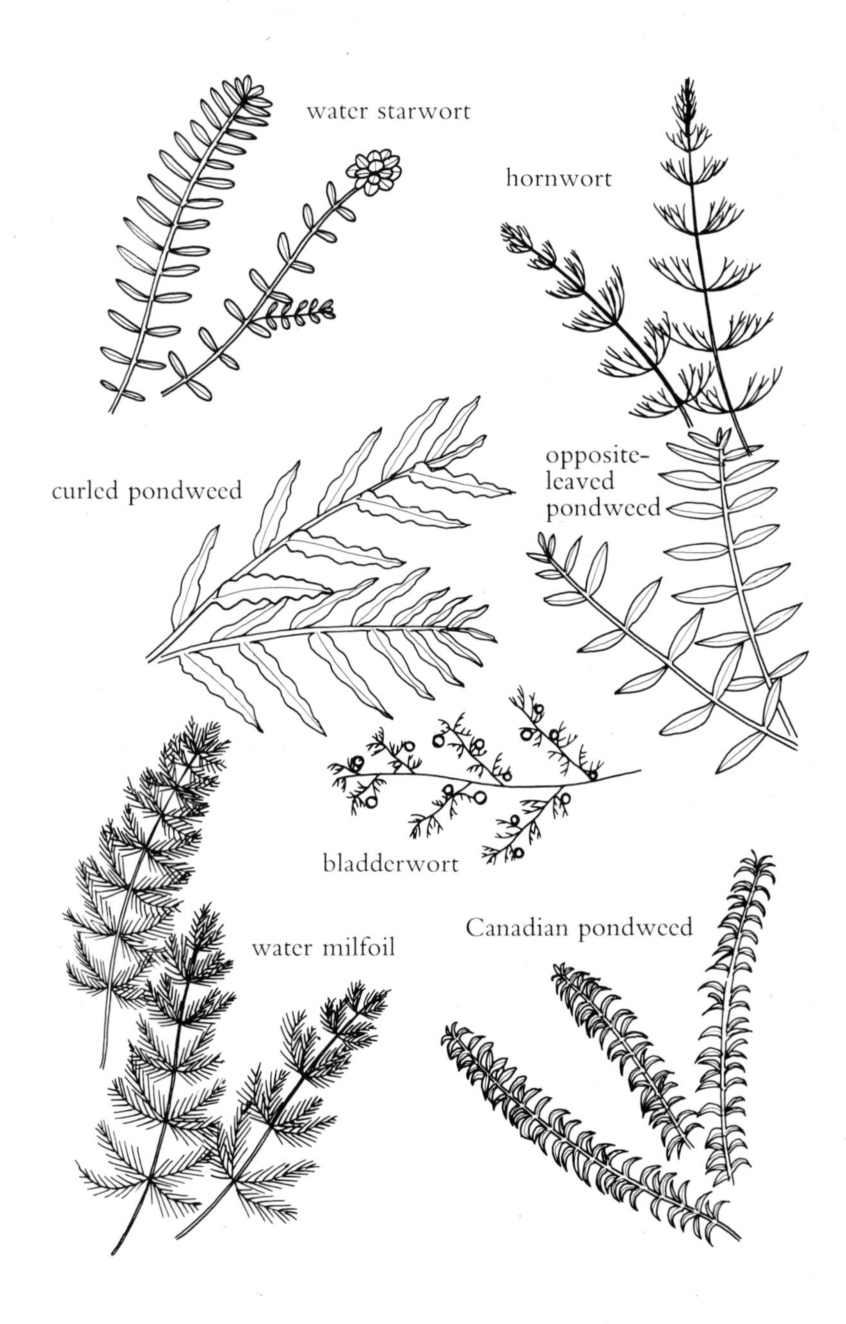

Figure 38 Some of the commonest aquatic plants.

above surface and when fertilised, bend under the water releasing fruit.

Starworts Another group of species amongst aquatic plants. Many species and many varieties within those species but all can be recognised by terminal star-shaped rosettes of leaves which give them their name. Often found in association with **water milfoils** and **hornworts**. Both are long, trailing plants with fine leaves, but those of milfoils are dissected like feathers. Both have rhizoids for attachment to bottom, but no proper roots, water being taken up by whole body surface. Hornworts are so completely adapted to aquatic life that they even flower and pollinate underwater. Flowers are small and insignificant, hidden away at axes of leaves.

Algae

Although too small to be seen by naked eye, algae are present throughout water column. Not only do algae supplement oxygen levels produced by larger plants, they also provide important food source for animals since they are at base of food web.

Occasionally during summer, algae increase so dramatically because of the sunlight that they cloud the water, turning it into a green 'soup'. These blooms often occur as a result of agricultural fertilisers seeping into water and may comprise many species.

Blue-green algae Simplest algae. In addition to green chlorophyll, other pigments give them colour. These are present throughout their cells and not contained in special bodies called chloroplasts as in other plants. Many are single-celled; others form large masses or filaments which sometimes clog up ponds and tanks.

Green algae Occur in variety of forms. Single-celled *Chlorella* is found in water column, but may also live symbiotically with *Hydra* giving it its green colour. **Colonial algae** are also found in water column. *Volvox* is a species where cells form a hollow space which moves by beating of tiny whip-like hairs or flagella. Examined under microscope, smaller spheres or developing 'daughter' colonies will be visible. Some species of green algae form long filaments with cells joined end to end. Filaments may be branched as in *Cladophora* or unbranched as in *Spirogyra*. In the latter, chloroplasts lie in a spiral along length of filament giving *Spirogyra* its name.

Diatoms Single-celled algae. A curious group which live inside silica cases. This can be best likened to a petri-dish with two sides, one fitting neatly into the other. To reproduce the two halves separate, each growing new replica of the other. May be found floating in water column or in masses on the bottom which creates silicous or diatomatous 'ooze'.

Stoneworts Although much more specialised than single-celled algae, with complex reproduction methods, they are sometimes classified with them. May be found in slow streams. Often carry white encrustation of lime which helps support them.

Flagellates

In some ways flagellates are half-way between plants and animals. *Euglena viridis* is a typical example. Possesses chlorophyll like algae but has whip-like flagellum with which it moves vigorously. Can, however, survive adequately in the dark. No true plant could photosynthesise and survive without light and so, in the dark, *Euglena* presumably feeds entirely by simple absorption. Other flagellates are even more like animals and feed by engulfing food particles.

Animals

A brief mention must be made of the classification of the animal kingdom. The major taxonomic animal groups are called phyla and the Arthropoda and Mollusca are examples. The next subdivision is a class. The phylum Arthropoda, for example, is divided into, amongst others, the class Insecta and the class Crustacea. To simplify matters I have omitted some groups of obscure animals and emphasised the more obvious ones. I have also occasionally used rather un-scientific groupings such as microscopic organisms where this helps in their understanding. I have selected some representative animals from each group and given details of their appearance, reproduction methods, feeding and respiration as a basic guide.

Microscopic organisms

The smallest, and in many cases, the simplest pond organisms, may be overlooked because of their size and, indeed, often require the specialised use of a microscope to see them (see Chapter 6). It is however, useful for all pond dippers to be familiar with the wide range of microscopic organisms.

Bacteria and fungi

Bacteria and fungi are two vital components of the freshwater environment, acting as saprophytes, living off organic remains and preventing the water from becoming choked. As such, they are essential for recycling of nutrients. Minerals and organic nutrients are released into water as result of decay and become available to living plants and animals for growth. If no decay took place, nutrients would remain 'locked' in dead matter at bottom of pond.

Protozoa

Single-celled animals with many species occurring commonly

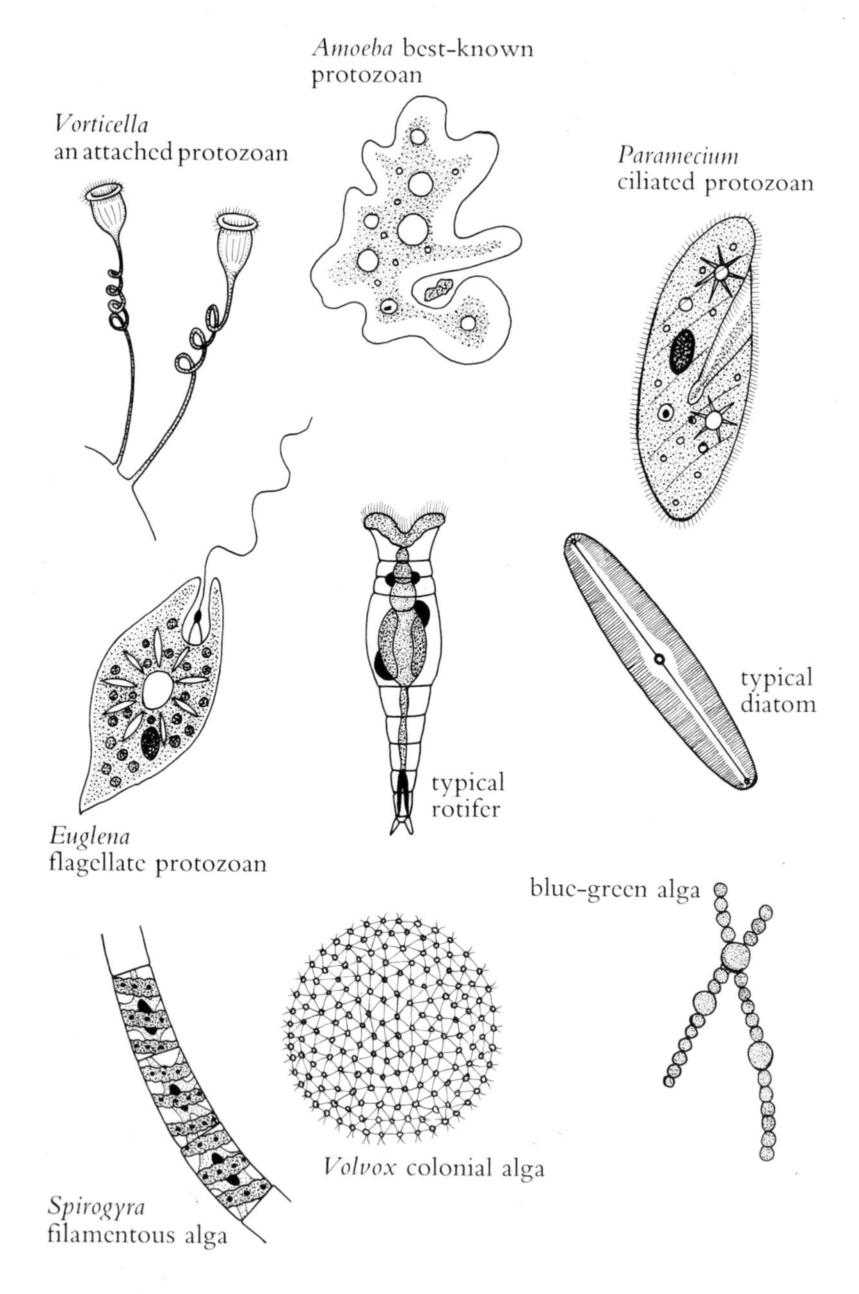

Figure 39 Although microscopic organisms are often the simplest pond organisms, they play a vital role in the ecology of a pond as a food source and in recycling nutrients.

throughout British freshwaters. Elevated to rank of sub-kingdom because of diversity of forms.

Amoeba Perhaps best-known protozoan. Exists as amorphous protoplasm, shape constantly changing as it oozes over substrate. Pushes out 'arms' or pseudopodia in direction of travel, the rest of the cell being dragged along behind. Plasticity of exterior allows it to engulf food particles, such as algal cells, into vacuoles in protoplasm and slowly digest them; remains are shed from vacuole. Vacuoles also help to remove excess water in essential process of osmo-regulation (see Chapter 5). Reproduces by asexual division and sexual conjugation.

Paramecium Coat of cilia or beating hairs enables it to move. More complex feeding method than *Amoeba*, food particles being channelled up oral groove and enclosed in vacuoles for digestion. Swims freely over pond bottom. Reproduction by simple division.

Vorticella Also typical of this bell-like genus. Makes no use of cilia for movement but attaches to plants in clusters.

Parasitic protozoa Sporozoan protozoa are true parasites producing large numbers of spores in their hosts. With their minute size you are unlikely to encounter them. *Opalina ranarum* A ciliate frequently found in rectum of common frog. Status as a parasite uncertain as it seems to do the frog no harm. It may simply be a commensal (sharing habitat) or have symbiotic relationship.

Rotifers

One of the most bizarre microscopic groups in appearance; also known as wheel-animalcules. Might be taken for large protozoa but are multi-cellular and belong to separate phylum. Name derives from ciliary ring or wheel surrounding mouth which creates current for feeding and movement. Generally cylindrical or cigar shaped although many variations. Transparent body through which well-developed gut and, in some species, eggs are visible. Males rare except at certain times of year so eggs produced parthenogenetically (without fertilisation). Generally swim freely although some attach to plants or stones and one species is a parasite of *Volvox*.

Sponges

More commonly associated with marine environment than freshwater but we have five representatives in our ponds and streams. Structure similar to marine sponges although not as large or spectacular. Encrust surfaces such as rocks and rooted plants, producing finger-like projections to increase surface area. Hollow chambers inside body connect to water by system of pores covering entire surface. Regular movement of flagella inside cells lining passages to and from pores, ensures a current of water passes

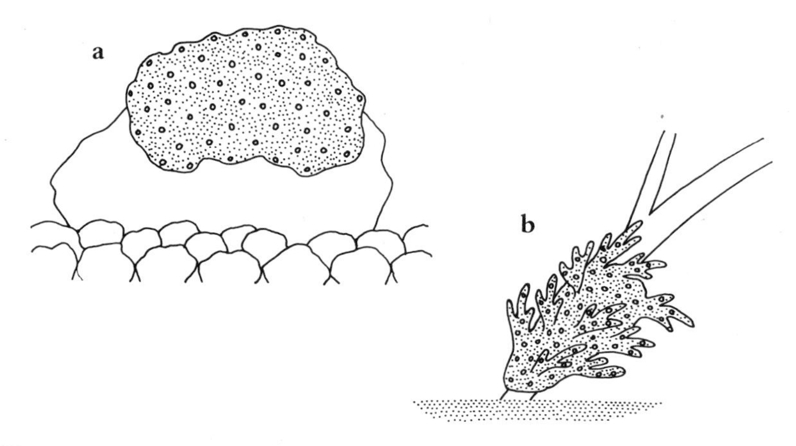

Figure 40 Two common freshwater sponges are the river sponge (**a**) which encrusts stones and the pond sponge (**b**) which encrusts the base of plants.

through to bring in food and remove waste products. 'Amoeboid' cells engulf and digest food particles. Silica spicules provide support for sponge body and prevent it from collapsing. Reproduce both sexually and asexually. Under favourable conditions sperm and eggs produced inside sponge. Currents caused by flagella enable transferral of sperm from one sponge to another and cross-fertilisation occurs. Resistant gemmules (cells from which new individuals develop) produced at asexual stage which overwinter.

Pond sponge and **river sponge** Two common freshwater sponges. Both form greeny-brown encrustations. The greater the exposure to light, the more intense the green becomes as symbiotic algae within sponge photosynthesise.

Coelenterates

Relatives of jellyfish and corals found in freshwater.

Hydra The most abundant and well-known coelenterate. Radially symmetrical like sea anemones, its marine relatives. Two layers of cells form hollow sac with opening at one end serving both as mouth and anus. This surrounded by ring of flexible tentacles which draw food in and allow water to pass out. Body contractions regulate the flow of water. Inside hollow sac 'amoeboid' cells engulf food particles and digest them; outside, stinging cells or cnidoblasts shoot out sticky thread or barb to trap passing prey. Toxins released simultaneously to paralyse prey while it is moved to mouth by tentacles.

Moves by 'looping the loop' rather like looper caterpillars of some moths. It may be important that *Hydra* moves from shade to light as, in common with freshwater sponges, it has symbiotic

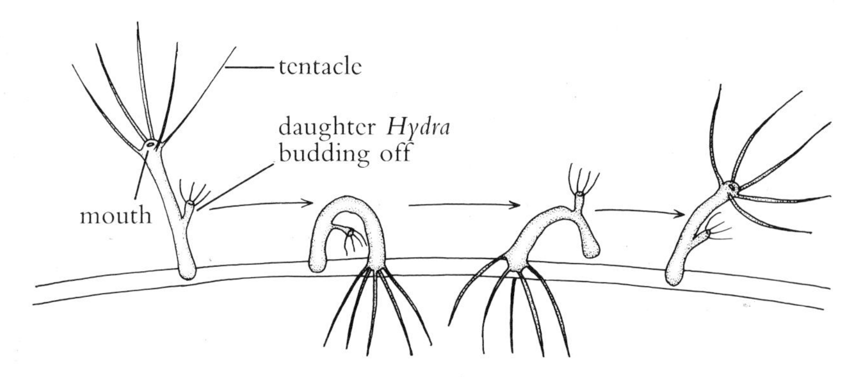

Figure 41 *Hydra* moves in a somersaulting action along plant stems by bending its body over and re-attaching its mouth further along the plant.

relationship with an alga giving it its green colour. Alga releases oxygen and uses up waste products including carbon dioxide necessary for photosynthesis.

Hermaphrodite (possesses both male and female organs) but only generates them under favourable conditions; general means of reproduction is asexual budding.

Worms

'Worm' is an artificial term applying to anything long, thin and wriggly! Comprise three phyla: platyhelminthes or flatworms; nematodes or roundworms and annelids or true worms. Leeches also belong to this last phylum, although superficially bearing little resemblance to true worms.

Flatworms (Platyhelminths)

Triclad turbellarians or planarians are the most frequently observed flatworms. Flattened dorso-ventrally with definite head and tail ends. Cilia on ventral surface enable them to glide with apparent ease over surface of rocks. Movement facilitated by 'slime' from head end; also cling to surface tension. Like leeches they have eye spots, number of which helps in identification; they lack suckers. Mainly carnivorous; digestion takes place in gut which divides into three, giving them name 'triclad'.

Of 12 species found in Britain, about four or five are widespread, in particular *Dendrocoelum lacteum* and *Polycelis nigra*. *Dendrocoelum* is white and gut and single pair of eyes can be easily seen. Large for a planarian (3 cm), it can be found both in running and standing waters. *Polycelis* only 1 cm and much darker, although colour variable. Found mainly in still waters.

Rhabdocoels Another group of turbellarians often overlooked because of small size, the largest being 2 mm. Some species make up

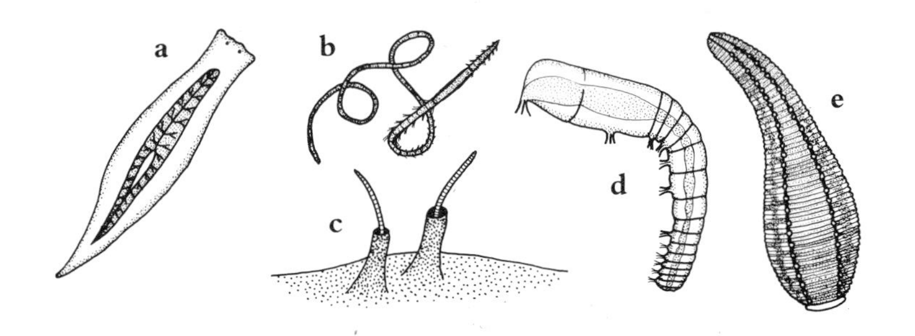

Figure 42 A variety of freshwater worms will be found in any pond. The planarian worm *Dendrocoelum lacteum* (**a**) glides over the surface of stones; *Tubifex* a true worm (**b**) lives in a tube in the mud and the tail end protrudes acting as a gill (**c**); *Chaetogaster*, another true worm (**d**); a leech is a flattened true worm and a sucker at either end aids movement (**e**).

for this by being brightly coloured. Glide by means of cilia; most species carnivorous. *Microstomum* however, is parasite of *Hydra* which absorbs stinging cells into its own skin for defence.

Trematodes You are unlikely to encounter this class of flatworms because adults are parasites living within host. Larval stages free-living but microscopic. **Liver fluke** is a notorious trematode which, as an adult, lives in bile-duct of sheep, cattle and sometimes humans.

Cestodes or **tapeworms** Parasites living within tissues of host. At some time during life-cycle parasite's offspring leaves to infect another host. Fish are particularly prone to infestation but usually debilitated rather than killed. Adult cestode found in vertebrates. Eels have adult *Bothriocephalus claviceps* tapeworm often in 'clumped' distribution common among parasite populations.

Cestodes *Schistocephalus solidus* and *Ligula intestinalis* infect sticklebacks and cyprinid fish respectively as larvae. Readily noticeable in infected fish since they cause large swellings on ventral surface. Life cycle of *L. intestinalis* is described in more detail in Chapter 5.

Roundworms (Nematodes)

Many species parasitic but others free-living amongst pond detritus. Occur in large numbers but are small and difficult to identify. Long, cylindrical and pointed at both ends. Lack longitudinal muscle so movements awkward and produced by s-shaped bends in body. To ensure survival eggs are very resistant and can tolerate both heat and desiccation. Eggs of *Ascaris lumbriocoides* (parasite of pigs) can survive for a year in formalin and still hatch.

True worms (Annelids)

Phylum divides into true worms or oligochaetes and leeches. Eight families of freshwater oligochaetes. All segmented, each segment being almost identical. Distinct head end with mouth and well developed gut and anus. Although hermaphrodite, many worms cross-fertilise and some mud-dwellers asexually 'bud-off' new worms.

Many freshwater annelids bear striking resemblance to **common earthworm**, which itself survives quite well in clean water, although not out of choice. *Tubifex* worms are confined to water and live in burrows in mud on bottom of ponds, especially where there is a low oxygen concentration. Tail ends are stuck out of burrow to act as gills and red haemoglobin in blood facilitates uptake of oxygen.

Hairs or setae on each segment help in identification but, more importantly, enable worm to grip substrate through which it is moving. Anyone who has watched a song thrush pulling an earthworm from its burrow can testify to the grip it has on the soil. Bristles not always evenly distributed and in *Nais* and *Chaetogaster* they occur in ventral bundles. Unlike *Tubifex*, these annelids are colourless.

Leeches Quite unlike true worms being considerably flattened dorso-ventrally. Segmented but possess two ventral suckers, one at either end, to grip substrate and move with characteristic 'looping' action. Bodies have great plasticity and when disturbed can contract into a blob. Swim actively and gracefully. Variable number of eye spots according to species. Pigment spots or chromatophores throughout skin and by contracting or expanding these, leech can alter its shade to suit surroundings.

Fourteen British species, all found in freshwater. Only a few are restricted to blood-sucking diet. Can penetrate skin with proboscis and set of jaws but only **medicinal leech** can bore through human skin. Other species are active carnivores catching small invertebrates. *Glossiphonia complanata* and *Erpobdella octoculata* catch small insect larvae, crustaceans and smaller snails, but only **horse leech** tackles larger prey. **Fish leech** Typical blood-feeder readily identifiable by prominent suckers. Occasionally seen attached to fins of fish such as minnow or trout.

Arthropods

Extremely important phylum of animals both in aquatic and terrestrial environments. Represented in freshwater by crustacea, insects and arachnids or spiders and mites. Characterised by and derive name from jointed limbs. Bodies segmented and, during course of evolution, segments have been lost or fused and appendages modified for sensory functions, feeding and movement.

Unlike vertebrates, arthropods have an external skeleton (exoskeleton) consisting of hard material called chitin and protein. Between segments, exoskeleton is soft allowing for flexibility; movement occurs by operation of internal muscles. Exoskeleton puts constraints on growth and the problem is overcome by moulting which may occur several times in the life cycle achieving step-by-step increase in size. Arthropods have well developed sense organs for touch, taste and vision. Insects and crustacea have compound eyes enabling them to be particularly aware of any movement.

Crustaceans

Primarily a marine class but well represented in our freshwaters. Exoskeleton often impregnated with calcium salts such as carbonate and phosphate; consequently crustaceans are found mainly in calcium rich or hard waters. In most species, exoskeleton simply covers body form but in some, such as ostracods, it may form shell similar to that of bivalved molluscs. Variable number of limbs for movement and feeding according to species and tendency towards fewer, more specialised types in advanced animals.

Respiration in smaller crustaceans may be by simple diffusion because of high surface area/volume ratio. Larger species have gills and often modified appendages to provide current of water.

Most larger crustaceans have separate males and females, eggs being laid and sometimes cared for by female. In smaller species males may be rare or unknown and parthenogenesis takes place. Young hatching from eggs may be miniature replicas of adult and will appear increasingly like them after each moult. Alternatively, eggs of some species hatch into larvae called nauplii which are quite dissimilar and metamorphose.

Five subclasses of British crustacea, some of which have only a few representatives while others have hundreds. Each subclass is described separately with its most characteristic representatives.

Branchiopods

The most primitive subclass. Three distinctive freshwater orders.

Fairy shrimp Quite large (up to 3 cm). Found in temporary pools, often in large numbers. Many segments, 11 of which have limbs enabling it to swim on its back and generating currents for respiration and feeding. Large stalked eyes (like crabs) are the most obvious characteristic since bodies transparent and almost invisible. Well adapted to live in temporary pools since eggs highly resistant to desiccation and drought; they will often appear suddenly in previously dry pools.

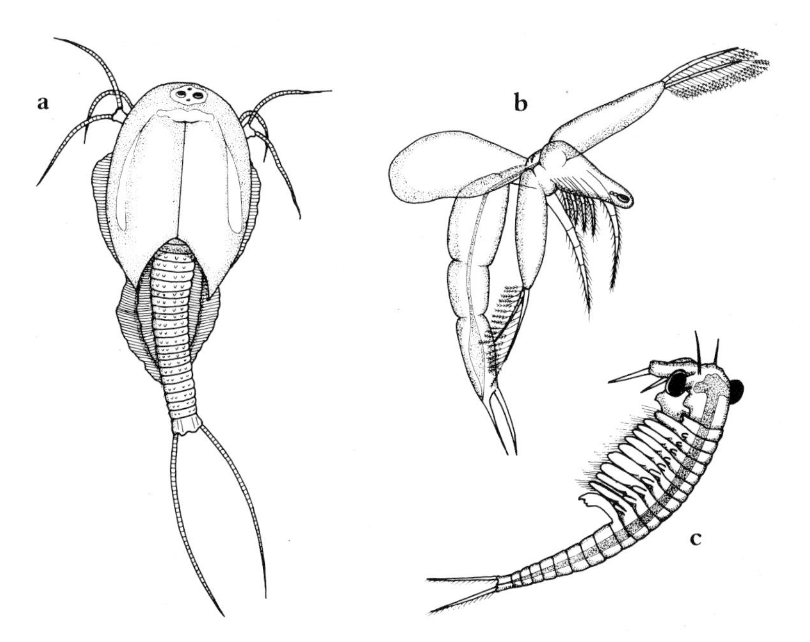

Figure 43 Three striking freshwater crustaceans. *Triops cancriformis* (**a**); *Leptodora kindtii* (**b**); *Chirocephalus diaphanus* (**c**).

Triops cancriformis A relative of fairy shrimp but rarer although distribution widespread. May be encountered in similar ephemeral pools. Striking appearance; it resembles miniature horseshoe crab with shell or carapace which covers head and thorax.

Water fleas More familiar animals of freshwater. *Daphnia* a common representative genus. Small, seldom larger than 2 mm with superficial resemblance to real fleas, which incidentally, are insects. Common name probably derives from hopping movement through water rather than appearance. This achieved by flicking of post-abdomen which projects beyond end of shell and encloses rest of body apart from head. Leaf-like limbs inside shell produce current of water which carries food and oxygen to mouth and gills. Bodies almost transparent and with a microscope you can see heart beating and eggs inside brood pouch which, when they hatch, look like small adults. Gut may appear green if water flea has been feeding on algae as is common.

Not all diplostracans resemble *Daphnia*. *Leptodora kindtii* although a relative of water flea has bizarre appearance with elongated body and long 'arms'. Formerly found only in glacial lakes in northern England but recently large numbers have been observed in some London reservoirs. Like *Daphnia* it has transparent body, but carnivorous diet. Can be found floating on upper waters of the lakes it frequents.

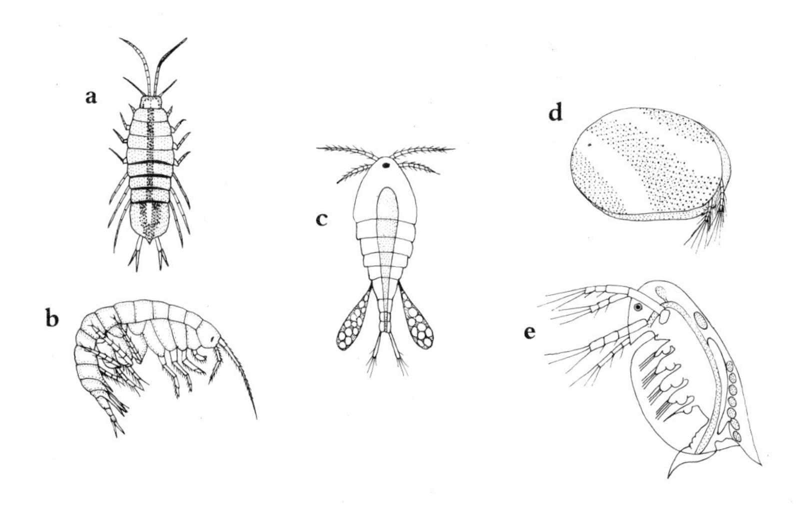

Figure 44 Freshwater crustaceans. Freshwater louse (**a**); freshwater shrimp (**b**); a copepod *Cyclops* (**c**); an ostracod (*Cypridopsis)* (**d**); a water flea (*Daphnia*) (**e**).

Ostracods

A sample of mud from the bottom of any pond will reveal a myriad of tiny, pea-shaped animals scuttling around. Ostracods belong to second sub-class of crustacea and the many species in British waters are rather difficult to distinguish from one another.

Cypridopsis vidua Characteristic with piebald mottling and bivalved shell. If disturbed, shell valves can be closed, but when open, two pairs of antennae and a pair of legs protrude. These waved vigorously to propel animal over surface of mud in search of food and they ingest food particles. Diet consists mainly of dead plants and animal matter and many hundreds may congregate around choice morsel. Eggs laid on or around food and produced parthenogenetically because males unknown in this species. Ostracods make interesting animals for the study of egg-laying in relation to density of adults and this is discussed in Chapter 6.

Copepods

Third subclass of crustacea. Representatives of this group seldom more than 1 mm long, but what they lack in size they make up for in numbers. Abound in their thousands in any pool and can often cloud water during summer months, especially since they often congregate near the edge and surface.

Cyclops Although small, such copepods can be recognised by pear-shaped appearance and single dark 'eye' which gives them their name. Move by jerky actions of limbs and antennae which give them a surprising thrust – they are impossible to catch with a pipette! Some individuals may appear to carry two dark sacs attached to abdomen; these are egg sacs.

Most copepods feed on planktonic matter, sorting specific size particles for consumption, but some such as *Ergasilus gibbus* are parasitic. The latter looks very similar to *Cyclops* but has two grasping claws which help it cling onto gills of eels – as if slime weren't sufficient! Others are parasitised themselves, for example by larval stages of tapeworms. The small worms are visible in body fluids of copepod under the microscope, but you will be lucky to find one, since average infection rate with tapeworm *Ligula intestinalis* is rather less than one in a thousand!

Branchiurans

Fish louse Our sole representative of subclass Branchiura, quite unlike crustacea described previously. Another parasitic crustacean, it is found, as name suggests, on fins and gills of many fish. Flattened and disc-shaped with two conspicuous suckers easily visible through transparent body. Indeed, these often the only things that can be seen when it is attached to host. Once on its fish, louse bores hole through skin with sharp proboscis and sucks blood.

Unlike many parasites, fish louse not fixed permanently to host but can swim around to find new host or improve its position. When it does so, fish seems very inquisitive and often tries to eat it, but unsuccessfully because of spines on the louse's back. There seems to be very little fish can do about this well adapted parasite. It doesn't appear to harm its host too much, but scars left by proboscis may allow fungal infection to enter.

Malacostracans

Last subclass of freshwater crustacea.

Water louse or **slater** Superficially resembles well-known **wood-louse** and is related to Isopods. Common in ponds and along slow-moving edges of streams, particularly where mud collects around roots of water plants. Finds food here consisting mainly of detritus; specialised mouth parts adapted for this. These compact with accessory appendages to help in feeding; food often held by front legs whilst chewed by mouth. Body flattened dorso-ventrally. Like many crustacea, possesses pigment cells or chromatophores in underlying cuticle which gives it its colour which can range from near black to colourless.

Freshwater shrimp Can also be found near edges of ponds and streams. Body flattened laterally so that it swims on its side. Abdominal legs generating currents for movement are in continual motion; also ensure flow of water over gills. If alarmed, shrimp can rapidly contract end of its abdomen and 'flick' through water like marine prawns and shrimps. Two individuals may often be seen apparently locked together. Here, male assists female, laden with eggs, to move. These will be looked after until they hatch.

Crayfish The only endemic freshwater decapod. Found in streams, especially chalky ones. Resembles small lobster with head and thorax fused to form cephalo-thorax, while abdomen remains segmented and flexible. First pair of walking legs modified to form large pincers, often of uneven size, which may be used in feeding or territorial disputes. These frequently lost or damaged but can be regenerated. Dextrous manipulation of food by appendages allows for rapid feeding. Claws probably more useful in defence and when competing for territories, although may sometimes be used to hold struggling prey. Same retiring habits as lobsters, preferring to live under stones or in holes on bank. Same rapid escape tactics as freshwater shrimp with violent flicking of abdomen. However, tip of abdomen broadened to form fan-like structure or telson which greatly increases thrust on water. During winter months, females carry eggs on underside of abdomen which remain with her until hatching the following spring.

Insects

Although largely a terrestrial class, insects are well represented in aquatic environment and are, to my mind, the most noticeable and fascinating group.

Insects are segmented and divided into three distinct regions: head, thorax and abdomen. Three pairs of legs on thorax, a character which readily distinguishes them from other arthropods. Both adults and larval stages are found in ponds and sometimes an insect spends its whole life in water. Many are only found in freshwater as larvae of which there are two basic types. Nymphs look just like miniature versions of adult, for example larvae of bugs. Beetle larvae in contrast, are completely different in form from adults and have resting or pupal stage during which metamorphosis occurs.

The sense of vision is particularly well developed. Many groups have compound eyes dominating head; in dragonflies they appear to wrap round it. Even dragonfly nymphs have large eyes which give them particularly good perception of movement.

A range of specialisations in breathing allows insects to obtain sufficient oxygen and expel carbon dioxide. Like terrestrial insects, aquatic ones breath via a complex system of tubes (tracheae) which branch and ramify throughout tissues. These open to the air via holes (spiracles) along the side of the body which allows gas

---

Figure 45 An adult greendrake mayfly resting after emergence. Huge swarms or 'hatches' can be seen by suitable rivers on warm spring evenings.

exchange to occur through all organs of the body. Flow may be increased by rhythmic pumping movements of body forcing air in and out. Often found in larval insects also. Aquatic insects often replenish air supply from water's surface. More specialised insects may trap an air bubble to keep themselves supplied – a form of physical gill – or may have proper gills with connections to air tubes.

It is amongst insects that true swimming first occurs as a means of movement in freshwater and to escape predation.

Aquatic insects are discussed below in systematic groups. All freshwater insect orders have common names which facilitates a grasp of their classification.

Springtails (Order Collembola)
Amongst the surface plants on any pond, little dark specks appear to jump along the surface. These are springtails or collembolans, and until they move, they appear to be inanimate objects. As name suggests, forked tail can be rapidly flexed to spring them into the air. Small and wingless. Unlike other insects, no tracheal system; instead breathe via skin and as a consequence lose water easily so confined to moist habitats.

Podura aquatica One of the commonest species. Almost black, it can be found amongst carpets of duckweed and water fern.

Dragonflies (Order Odonata)
A characteristic feature of ponds in summer. Two pairs of wings, strengthened by veins making them strong fliers. Order subdivided into Anisoptera or dragonflies and Zygoptera or damselflies. Two types can be distinguished by position of wings at rest. Damselflies are much smaller and fold wings over their backs in contrast to spread-out wings of dragonflies.

Dragonfly nymphs are aquatic and bear vague resemblance to adults except for lack of colour and wings. Both voracious carnivores, adults catching insects on the wing and nymphs eating any invertebrate small enough to catch. Nymphs have curious extension below the mouth (mask) which can be shot out to catch prey some distance away. Despite its appetite it may take several years for a nymph to grow large enough to emerge into an adult. Rectal gills facilitate oxygen uptake and nymphs can sometimes be seen rhythmically expanding and contracting abdomen to cause flow of freshwater over gills.

Emperor Largest British dragonfly. Blue abdomen with black line down centre. Common on any large pool and a very fast flier,

---

Figure 46 A water stick-insect in a characteristic pose waiting for prey. Its sharp rostrum is inserted into the prey while the juices are sucked out.

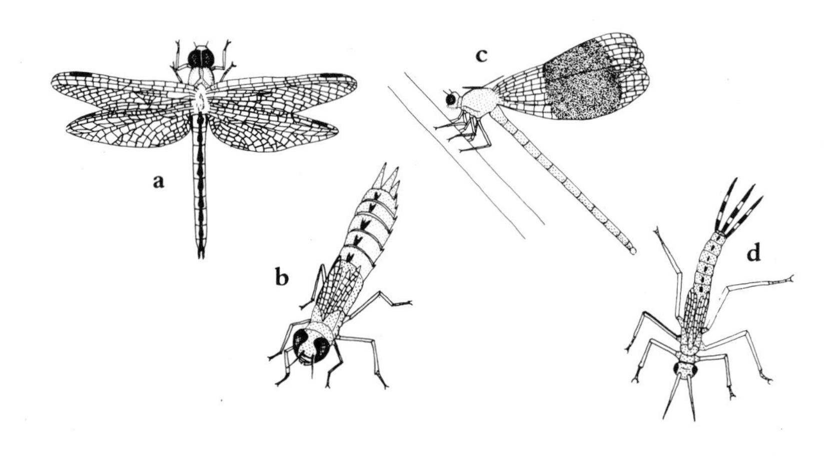

Figure 47 Some common dragonflies and damselflies. The wings of the adult emperor dragonfly (**a**) are held flat at rest. The emperor dragonfly nymph (**b**) is an active species which is found amongst pond weeds. The wings of the adult banded agrion damselfly (**c**) are held closed behind the body at rest. The distinguishing features of the banded agrion nymph (**d**) are the three caudal lamellae or gills at the tail end.

patrolling its own patch or territory. Nymph also large, often over 5 cm, with characteristically large round, flattened eyes. Can be found amongst weed, often near pond surface. If disturbed it can escape rapidly by its own jet-propulsion by shooting water out of anus.

Broad-bodied libellula Nymph lives buried in mud at bottom of pond. Tip of abdomen and eyes are all that protrude as it waits patiently for passing prey. Well adapted for life in mud as it is flattened and so does not sink. Protruding abdomen allows rectal gills to function. The large number of hairs on its back pick up particles of silt and give additional camouflage. Adults are striking with broad, flat abdomens; male blue and female orange.

Damselflies appear dainty in contrast to sturdy dragonflies.

Blue-tailed damselfly Particularly delightful species. Bright blue penultimate segment on abdomen gives impression that it is carrying a blue tail-light and is often its most noticeable feature. Nymph can be found amongst the pond weeds. Its three gills (caudal lamellae) projecting from tip of abdomen contain an intricate network of tracheae to absorb oxygen. Shape of these three flattened plates is crucial in identification; in this species they are pointed. (Caudal lamellae can be lost with no ill effect, but probably enable survival in lower oxygen concentrations.)

Banded agrion Our largest damselfly which really lives up to its Latin name *Agrion splendens*. Wings have metallic appearance and deep blue patch in middle. Although sluggish fliers, wings catch the light making them shimmer. Courtship dances are performed over streams with up to a dozen males flying together – an unforgettable sight. Nymphs found at bottom of streams, often half buried in mud. Nymphs of both banded agrion and its close relative, demoiselle agrion have first segment of antennae greatly enlarged giving them horned appearance.

Stoneflies (Order Plecoptera)
Rather dull creatures which spend much of their adult lives hidden amongst vegetation by water and are easily overlooked. Poor fliers which soon come to rest if disturbed. Two pairs of wings; relatively large and membranous; lie flat on body when insect at rest. In this position prominent pair of antennae and pair of tail appendages or cerci protruding from rear end are visible.

Isoperla Common and widespread genus found in streams as are most stoneflies. Flowing water maintains good aeration, perhaps one reason why stonefly nymphs seldom found in still waters. Very sensitive to changes in oxygen levels and one of first groups of freshwater invertebrates to disappear if a stream becomes polluted. Nymphs spend several years in water hiding under stones and share adult characteristic of the cerci.

Mayflies (Order Ephemeroptera)
One of the glories of British waters. Adults have very short lifespan during which they do not feed. To ensure successful mating, 'hatch' or mass emergence is synchronised. A few hours before sunset males form large swarms and fly up and down the same spot of river. Females fly into these, quickest males grab them and mating occurs in mid-air. Females then fly off to lay their eggs, either on water surface or underneath, depending on species. Males, however, die almost immediately and drop onto water, sending fish, notably trout, into a feeding frenzy. Mayflies therefore have a great attraction to fishermen who imitate dead flies with artificial ones. It is relatively easy to watch mayfly hatches in the wild. Pick a mild, calm May evening and walk along the banks of a stream. Although they have their favoured areas, you should be lucky. You may have to visit on several evenings as the hatch is unpredictable.

Nymphs may spend up to three years in water depending on species. Gills along side of abdomen are formed around extensions of tracheal system; three cerci protrude from this also, distinguishing them from stonefly nymphs. Nymphs moult many times but three distinct stages. Size of wing buds increases with each moult until in last stage, or sub-imago, wings are fully formed and

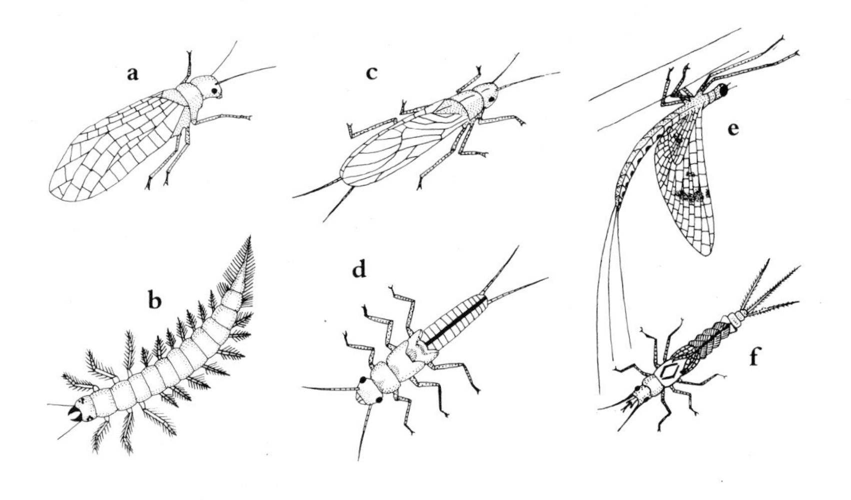

Figure 48 A selection of freshwater insects; the adults are terrestrial and the larvae or nymphs aquatic. The adult alderfly (**a**) can be found resting on plants by ponds and rivers. The alderfly larva (**b**) is predatory; its gills are on the side of the abdomen. The adult stonefly (**c**) can be found amongst waterside vegetation; it is a reluctant flyer. The stonefly nymph (**d**) can be found under stones in clean rivers. Large numbers of the adult greendrake mayfly (**e**) hatch simultaneously. The greendrake mayfly nymph (**f**) lives amongst the sediment at the bottom of slow moving rivers; its gills are on the abdomen.

insect can fly. This stage unique among insects but lasts only a short time, before final moult produces adult insect.

Greendrake Common species found in early June toward edges of slow moving streams. Nymph burrows in soft mud but others are adapted for crawling, clinging to stones or swimming to escape predation.

Pond olive Has highly active swimming nymphs, common in still water, particularly small ponds.

Bugs (Order Hemiptera)

Numerous in our freshwaters and individuals of various species can be found in all months. Identifiable by piercing mouthparts in form of beak or rostrum on front of head. This necessitates liquid diet which may be plant or animal material.

Most aquatic bugs spend their whole lives in water, although winged species sometimes fly from pond to pond. Water cricket and pond skaters have wingless adults. Winged forms may be found occasionally; this may be connected with the drying-up of habitat, the wings allowing them to move farther afield.

Water boatman A familiar sight at pond surface. *Notonecta glauca* the commonest species. Can be recognised by large size (up to 2 cm) and habit of swimming upside down. Remains floating at surface because of air trapped by hairs on its abdomen, a physical 'gill'. This bubble used as air supply for respiration but also gives it buoyancy. If alarmed, it will swim rapidly downwards and hold onto plants at pond bottom. Third pair of legs is longer than others and fringed with hairs making an effective paddle. A ferocious carnivore, often tackling large animals such as tadpoles with rostrum. Can also pierce human skin, so take care if handling one. The bite is particularly painful as it injects toxin and enzymes which would normally paralyse and digest a smaller animal.

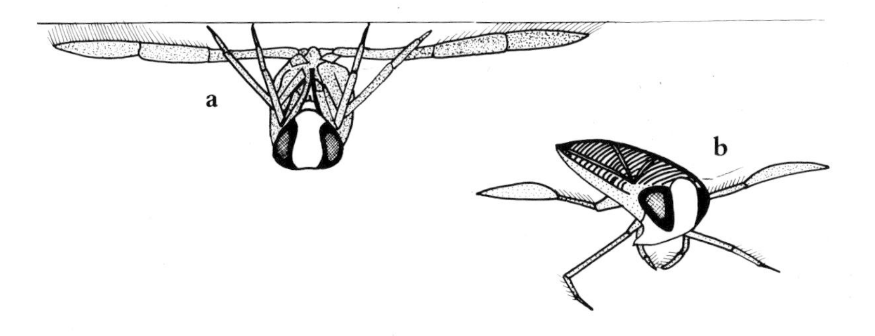

Figure 49 The water boatman (**a**) often clings to the surface film. It is an active carnivore with piercing mouthparts. The lesser water boatman (**b**) swims the right way up and feeds on detritus by 'hoovering' it up.

Lesser water-boatman Could be confused with water boatman but distinguishable by its small size and habit of swimming the right way up. Air held in depression on ventral surface of abdomen. (Nymphs breathe through skin.) Third pair of legs again used for paddling but, unlike its larger relative, it feeds on detritus sucked up with its rostrum like a vacuum cleaner. In contrast to *Notonecta* it makes an ideal addition to an indoor tank and will help to keep it clean. Both move very rapidly if alarmed, but this is not typical of all bugs.

Saucer bug Found in most still waters. Can grow to 1 or 2 cm and is distinctly flattened and oval, hence the name. Could be confused with two other bugs, *Plea leachii* and *Aphelocheirus montandoni* but these are both much smaller, rarely exceeding 3 or 4 mm. Unlike saucer bug, *Plea* swims upside down and congregates in vast numbers amongst pondweeds. *Aphelocheirus* can be found under stones in rivers. Adults lack wings.

Water scorpion At first sight this may not be recognised as an animal let alone a water bug. Body so compressed and oval that it looks just like a dead leaf. Camouflage increased by long breathing siphon from abdomen which could be the leaf petiole. Lies motionless in water waiting for unsuspecting prey and not until it is disturbed is it readily noticeable, making frantic movements with its legs. If caught unbeknown in net and taken out of water, it will often remain motionless, making detection very difficult.

First pair of legs at rear used to grip prey, like forelegs of praying mantids. Hooks on tips of legs facilitate this and, if successful, legs draw prey towards head impaling it on long proboscis or rostrum. Legs are then held in striking position of prayer. After digestion contents are sucked out.

Water stick-insect Very similar in appearance to its terrestrial counterpart. True stick-insects are in no way related to bugs. Water stick-insect carnivorous with proboscis like that of water scorpion. Both water scorpion and water stick-insect often plagued with parasitic mites, larvae of species such as *Hydrarachna*.

Water measurer One of the bugs specialising in life on the surface film. It is the least agile, preferring sides of ponds and slow moving streams. It has a coating of water-repelling hairs.

Pond skaters These familiar insects are more daring and will dash unhesitatingly to any insect that falls into water and is stranded on surface film. Its struggling motions send out ripples and it is these vibrations that the skaters detect with their long, sensitive legs. Pond skaters are often found in company of **water cricket**.

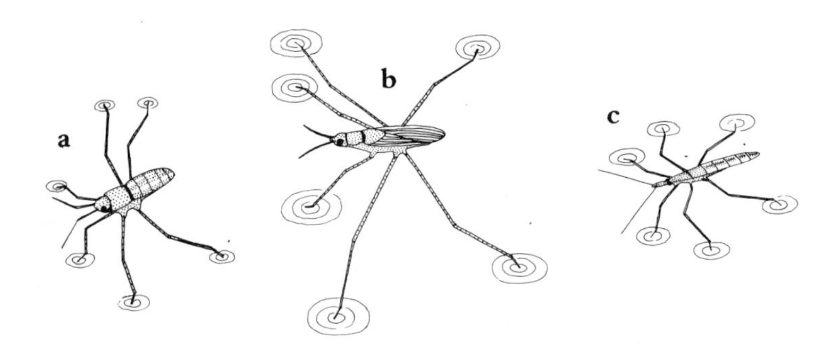

Figure 50 Several freshwater bugs live on the surface film. The water cricket (**a**) can frequently be seen on ponds and at the edges of slow moving streams. The pond skater (**b**) is a fast moving species which is found on most ponds. The water measurer (**c**) is another common species.

Alderflies (Order Neuroptera)

Unlike the nymphs of bugs which are like miniature adults, alderflies' life history involves egg, larva very different from adult, pupa and adult itself. Larva could be confused with that of a beetle, as both have long, tapering shape. However, alderfly larvae have a 'pointed' tail whereas beetle species have two cerci. Seven pairs of lateral abdominal gills behind legs are another distinguishing feature.

Larvae leave water to pupate underground and hatch into adults somewhat similar to stoneflies but without the two cerci. *Sialis lutaria* the commonest of our two species. Not a powerful flier. Can be seen during May and June.

Beetles (Order Coleoptera)

Like alderflies, water beetles have four stages in life cycle. In appearance very similar to terrestrial beetles. Adults have first of their two pairs of wings hardened and shaped to form coverings or elytra which protect hind wings and abdomen when at rest. The many species have a wide range of diets from herbivore to carnivore. Mouthparts basically similar, being adapted for biting. Larvae also have biting mouthparts but are invariably carnivores and may be active swimmers or crawlers in pursuit of prey. Pupation frequently takes place away from water and larva often prepares a special chamber underground or puparium for added security.

Beetles are primarily adapted to life on land. Larvae, although confined to water, breathe air from surface by sticking abdomens above water, the spiracles connecting to tracheae. Adult water beetles have solved problem of breathing underwater by simply

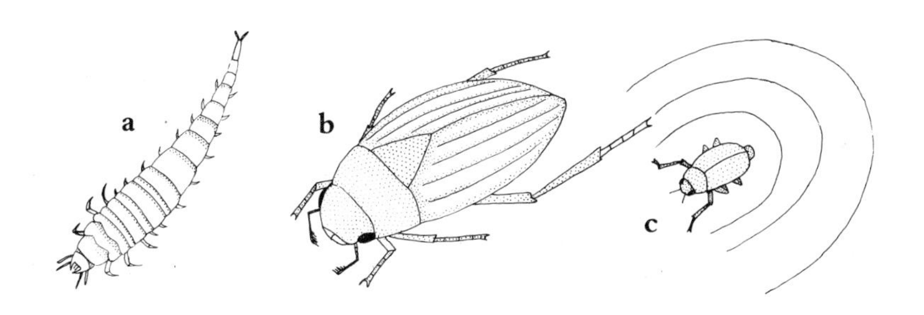

Figure 51 Some common aquatic beetles. The larva of the silver water beetle (**a**); adult silver water beetle (**b**); adult whirlygig beetle (**c**).

taking air supply down with them. They trap thin bubble of air under elytra and sets of velvet-like hairs help retain it. Air is in direct contact with spiracles of thorax and abdomen. This bubble also acts as a physical or plastron 'gill'. If oxygen concentration in plastron 'gill' drops below that of surrounding water, then oxygen will diffuse in. Conversely, carbon dioxide will diffuse out if concentration rises too high. In this way air supply lasts much longer. From time to time of course, beetle has to replenish air. Great diving beetle does so by sticking tip of abdomen through surface of water. Hydrophilidae such as *Helophorus* are a small group of beetles that are slow swimmers and can often be seen on surface. Their method of replenishing air is ingenious. Last four segments of antennae are grooved and form a tube when pressed against a groove on head which connects with air under elytra.

Five main families of beetle found in freshwater.

Great diving beetle (Family Dytiscidae) Elegant adult is large and greeny-black with orange margin. Despite regal appearance, it is a fierce carnivore and using its paddle-like hind legs, actively pursues prey as large as tadpoles, minnows, and sticklebacks which it will eat remarkably quickly. Males can be distinguished from females by suckers on tarsal segments of front pair of legs which are used to grasp female during mating. Hairs on either side of hind legs form paddles enabling it to swim rapidly. Larvae just as ferocious as adults, if not more so, and have long mandibles with which they pierce prey. Digestive enzymes forced into prey through canals in mandibles which connect with pharynx. Resulting soup is then sucked up by larva – a truly horrible way to go!

Acilius sulcatus Smaller with beautiful gold markings along grooves or sulcations on elytra. Larvae and adults found commonly in most ponds, larvae being distinguished by elongated first thoracic segment.

Screech beetle (Family Hygrobiidae) Sound it makes is produced by rubbing wing case against tip of abdomen. In contrast to great diving beetle's streamlined appearance, screech beetle is well rounded and has pronounced neck with head sticking out. Larva too looks very different with large head and thorax and tapering abdomen.

Family Haliplidae Poor swimmers and their floundering movements might be mistaken for those of a drowning insect. Prefer to live and feed amongst filamentous algae which they pierce with mouthparts to suck out contents.

Whirlygig beetles (Family Gyrinidae) Superb swimmers, often collecting in large numbers on pond surface. They can fly but their response to alarm is to dive using third pair of legs as paddles. Extremely sensitive to movement and will disappear in an instant at your approach. Only sign of their presence will be a small, silvery bubble of air, protruding from tip of abdomen. This bubble is their air supply, replenished from time to time at surface. Larvae rather different from those of other beetles, being long and thin with tracheal gills along sides. Could possibly be confused with larvae of alderflies or caddisflies. When ready to pupate, larva crawls up water plants and makes cocoon between the leaves, above water.

Silver diving-beetle (Family Hydrophilidae) Our largest water beetle; may reach 5 cm in length. Film of air trapped by hairs under abdomen gives it a silvery appearance. Has become rare recently partly as a result of collection for stocking tanks. Its popularity is due to its elegant appearance and also because adult beetle is herbivorous and not as destructive as some of its relatives. Larva, however, is carnivorous and has sizeable mandibles with which it tackles prey, in particular water snails.

Moths (Order Lepidoptera)

One would perhaps not immediately associate moths with freshwater. Many species, such as the **drinker moth** frequent wet habitats, the larvae feeding on lush grass. However, a few species, particularly **China mark moths** (family Pyralidae) have aquatic larvae which feed on water plants. In some species eggs are laid under lily leaves and young larvae actually live within leaf tissue. As larvae grow larger they construct portable case from leaves which remains filled with air, trapped by hairs. Larvae then crawl around under water's surface and eat characteristic oval-shaped holes in leaves.

Caddisflies (Order Trichoptera)

Adult caddisflies have a similar appearance to moths but belong to a separate order. Instead of scales, hairs cover surface of wings and at rest these are held in a 'roof' over the abdomen. Despite size of wings, they are poor fliers, only reluctantly taking to the wing.

Eggs are deposited on water surface or on water plants, either

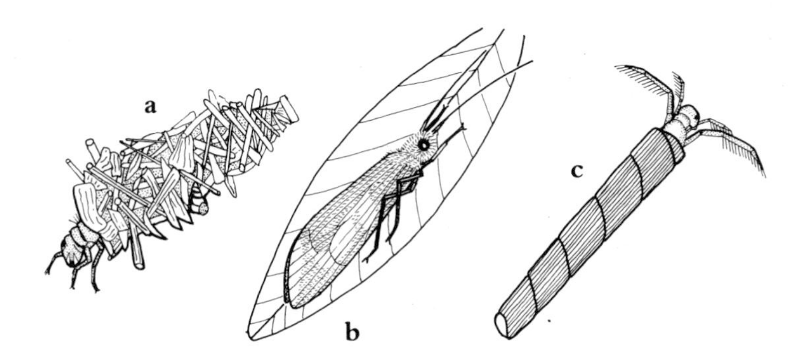

Figure 52 Caddisfly larvae manufacture a variety of protective cases which remain all their lives. The larva of *Limnephilus* (**a**), a crawling species; the case is made from twigs and leaves; the adult caddisfly (**b**) looks like a moth except for the wings which are hairy; the larva of *Triaenodes* (**c**), a swimming species; the case is made from rolled up leaves.

above or below surface. Hatch into endearing larvae, renowned for manufacturing protective case which they carry all their lives. Identification of many species can be determined by the cases, as each is constructed differently. Some larvae use plant material for these whilst others use grains of sand or even snail shells. As caddis larvae are generally slow moving, crawling around on pond bottom and amongst vegetation, the case helps protect them from predators. Not all species have larvae with cases; some are free-living making silk tunnels and nets.

Since case-forming caddis larvae are protected, only head, thorax and first few abdominal segments need be hardened. Rest of abdomen is softer, with lateral gills formed around extensions of tracheal system and a pair of hooks at tip which grip inside of case.

Larva of caddisfly *Triaenodes* builds lightweight case of strips of leaves. Long case gives essential streamlining since this species swims actively, the third pair of legs being long and paddle-like. Species of *Limnephilus* construct a sturdy, heavy case out of twigs and larger pieces of plant. Larva generally crawls around bottom of ponds and lakes, although occasionally can be found floating on surface with bubble of air trapped inside case.

Flies (Order Diptera)

True flies have only two wings, second pair having become stabilising organs (halteres). All adult flies are terrestrial but many have aquatic larvae. Adults are liquid feeders with a proboscis designed for sucking up nectar, rotting fluids or blood. Well developed eyes, but these characters generally lacking in larvae,

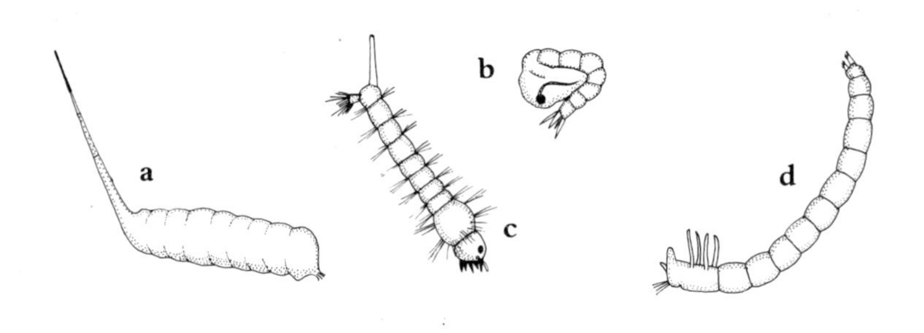

Figure 53 Freshwater fly larvae and pupae. The larva of the rat-tailed maggot (**a**) is found most often in polluted waters; the mosquito pupa (**b**) lives at surface of ponds and rain butts; mosquito larvae (**c**) will be found alongside the pupae; the red colour of the chironomid larva (**d**) is due to a blood pigment; it is found in the mud of the pond bottom.

which are mostly degenerate and maggot-like. Exceptions to this are flies such as mosquitoes, larvae of which are more highly developed, floating at surface of water. All fly larvae possess internal air-filled tracheal system but spiracles may be reduced, modified or absent. Some have anal 'gills' but these probably have more important role in absorption of dissolved salts. Their general body surface is probably more important in exchange of gases. Larvae are mostly filter feeders or detrivores.

Daddy-long-legs Adults of tipulid family. Familiar lumbering flight which sometimes causes mild irritation when they fly into your face. Other biting flies are more dangerous. Larvae are **leatherjackets**. Many species are found in freshwater. Larvae are fleshy with poorly developed head which can be retracted into body. Common amongst decaying plant material at bottom of ponds and streams.

Culicid mosquitoes Transmit malaria in tropical countries. Only the females bite and in many species a blood meal is essential for development of eggs. Eggs are laid in almost any water body including rainbutts and hatch into active swimming larvae. Bristles on mouthparts enable them to filter-feed for planktonic life at surface. Both larvae and pupae are found here and, if disturbed, can swim to escape danger. Breathe aerial oxygen and have ring arrangement of hairs on top of abdomen. These fan out on surface tension when abdomen touches it, allowing contact between abdominal spiracles and air.

Blackflies Simulidae Notorious for painful bite. Prefer to live by flowing water; larvae and pupal cocoons found attached to stones.

Chironomids Larvae are small, red 'worm-like' creatures belonging to genus *Chironomus*, found in mud at bottom of pond. Red colour is a respiratory pigment (erythromycin) similar to haemoglobin enabling larvae to take up extra oxygen and live in water with poor aeration. Some species live in fragile mud tubes which they hold on to with thoracic and abdominal pro-legs whilst undulating body to generate water current to anal gills. Chironomids are important food items for young fish and small invertebrates, although in many species of fly, larvae have evolved to survive in waters incapable of supporting predators such as fish, and their presence indicates pollution.

Clegs or **horse flies** Common in New Forest and Dartmoor. As with mosquitoes, only the females bite. Larvae have characteristic tapering form with thickened rings along body. Can be found at edges of streams, particularly amongst moss and organic debris. Some feed on debris whilst others predate small invertebrates.

Rat-tailed maggot Larva of drone fly which frequents the most stagnant and polluted waters. No head, the most notable feature being a long, extensible respiratory siphon protruding from tail end which connects anal gills with aerial oxygen. All goes to give a truly bizarre, if not repulsive appearance.

Groundhoppers (Order Orthoptera)

A section on Orthoptera may seem out of place in a book on pond life. However, the group not only includes grasshoppers and crickets, but also a ·largely unnoticed group of animals: groundhoppers or Tetrigidae. These tiny insects, between 1 and 2 cm long, look like miniature, armour-plated grasshoppers. Commonest species *Tetrix undulata* prefers dry habitats but two other species, **slender** and **Cepero's groundhopper** have particular liking for old marl pits. This habitat is found all along south coast, especially in the New Forest where clay has been worked and spoil tips left for many years. These fill in with water and gradually become colonised. Around edges *Sphagnum* moss plentiful, upon which groundhoppers feed. It is their escape mechanism from terrestrial predators which links them with freshwater. When alarmed they jump like grasshoppers, but do so toward water and can submerge and swim for some time.

Spiders and mites

This distinctive class of arthropods is characterised by the four pairs of legs attached to thorax. Like insects, body of a spider can be divided into three distinct regions, head, thorax and abdomen. In contrast, abdominal segments fused and separated from thorax by narrow waist. Mites have all the body segments fused into one, giving them the appearance of minute spheres with legs.

Water spider The only true aquatic spider in Britain. This delightful animal is common in many ponds and its silvery form can sometimes be clearly observed from above the water surface. Sheen is due to layer of air surrounding abdomen, trapped there by coating of hairs. This forms part of its air supply, although it also constructs air-bells amongst the water weeds in which it lives, the air being trapped by silk webs. Eggs are laid within this diving bell and guarded and fed by female, a rare example of parental care amongst pond invertebrates. Other diving bells are used for courtship, mating and overwintering. Long legs make the water spider highly suited to clambering amongst pond weeds.

Swamp spider A species exclusively found on or beside water. As name suggests, it likes swampy places, especially where pond margins are covered with *Sphagnum* moss. Our bulkiest species it is a beautifully coloured chocolate brown with yellow stripes. Whilst waiting for prey it rests at pond edge with first pair of legs feeling for vibrations on water surface which might indicate a drowning insect. When it senses this, it will rush across the surface to catch prey. Legs are very hairy so that it can walk on surface tension. If alarmed, it climbs under water surface and escapes by climbing down under water plants. Like water spider, a film of air trapped around abdomen gives it a silvery appearance.

Mites Extremely small, rarely exceeding 2 or 3 mm. Often brightly coloured (particularly red) and so not difficult to spot as they scuttle along, their legs moving frantically. Genus *Hydrarachna* is particularly common. Adults bright red. Larvae are parasites of freshwater invertebrates and rather dull in colour. These are found on water scorpion and water stick-insect. When fully developed, larvae drop off host and become free-living adults.

Snails (Molluscs)

A very diverse phylum both in size and body shape. Represented in British waters by bivalved molluscs and snails. These very dissimilar in anatomy although they do have the shell in common.

Protective shell is essential since the body of a mollusc is very soft. Body unsegmented but divided into various regions such as hepatopancreas (digestive organ), reproductive glands and foot. Latter region is a finger-like projection in bivalves which can lever body along in a rather clumsy fashion and anchors shell in mud and silt in which snail lives. True snails or gastropods have a flattened foot, used for gliding over surfaces by a combination of muscular and ciliary action. Edges of body are extended and flattened to form layer called the mantle. This tissue secretes the shell. Although shell is inert, it is produced by a biological process and grows continually.

Shell comprises three layers: mother of pearl or nacre, calcium carbonate and horny layer of protein on outside. Nacre is secreted by whole of mantle and, in addition to forming the shell in bivalves, it is also used to encapsulate foreign bodies and parasites. Hence pearls are formed in oysters and smaller pearls sometimes found in freshwater mussels. Calcium carbonate layer is a rigid structure of crystals which lies perpendicular to shell. This is covered by protein layer, although in older animals this may be chipped off, something particularly noticeable in larger freshwater mussels.

The two outer layers of shell are laid down at edge of mantle. Growth occurs at faster rate in summer and growth rings are visible, particularly in bivalves. Snail shell is essentially a thin cone twisted either in right-handed direction (dextral) or left-handed one (sinistral). This process (torsion) occurs during development of snail. In the gastropod's shell, growth occurs by the lengthening and enlargement of mouth of this cone.

A mollusc's shell is not a temporary home like that of caddis larvae or hermit crab but an integral part of its body attached to it by muscles. Without its shell the snail would die. This dependence on a shell, composed largely of calcium carbonate, means that snails are

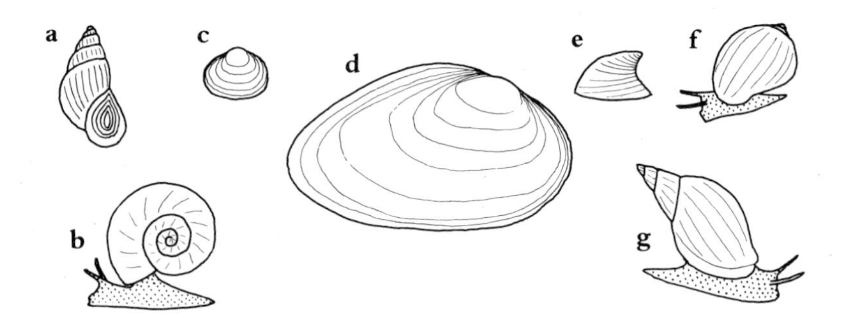

Figure 54 Freshwater molluscs. Common bithynia with operculum (**a**); ramshorn (**b**); orb shell (**c**); swan mussel (**d**); river limpet (**e**); wandering snail (**f**); great pond snail (**g**).

largely restricted to hard or calcium rich waters. No snails are found in acid heathland pools where the water is very soft.

Gastropods or true snails

Literal meaning of stomach-foot refers to the most obvious soft part of snail's body – the foot which contains many of the vital organs. Gastropods are divided into two groups: pulmonates and operculates.

Pulmonates

As name suggests, these have a lung or chamber within shell surrounded by mantle tissue. This connects to air via tube with an opening or pneumostome. Periodically pulmonates visit surface to replenish air supply which provides a good opportunity to watch them closely.

Pulmonates are hermaphrodite, having both male and female organs. Quite capable of self-fertilisation but, given the opportunity, will mate with another snail with acting male climbing on shell of the other. To complicate matters further, other snails may attempt to mate with 'male' snail in a partnership already taking place. Here third snail fertilises female part of 'male' snail. Resulting eggs are embedded in familiar strips or blobs of jelly which can be found under lily leaves.

Great pond snail One of our largest species. Its pneumostome can be seen easily especially as it seems to spend a lot of time at the surface. Air in lung gives water snail buoyancy as well as oxygen supply. Even the heavy great pond snail can float well and it often glides upside down, holding onto surface tension with its foot. If disturbed pulmonate snails can quickly expel air from the lung and sink to the bottom. Pneumostome closes at this point to prevent any water getting in. Great pond snail has many relatives in freshwater.

Wandering snail perhaps the commonest water snail, often found some distance from water. A tiny relative is **dwarf pond snail** which prefers water margins and is often found in poach marks made by cattle when they come down to the water. As an intermediate host of liver fluke, it is unpopular with farmers.

Shells of limnaed snails form spires. Those of ramshorn snails or planorbids are flattened resembling a catherine-wheel. 'Lungs' in planorbid snails may be full of air or water and haemoglobin stores and carries oxygen. May also have gill to aid respiration. There are many species of planorbid.

Ramshorn One of largest and most common planorbids. Planorbid and limnaed snails feed by grazing algae off surface of plants and stones with rasping tongue or radula as at least part of their diet. Also ingest partly rotted plant material from pond bottom.

Limpets on the seashore are familiar to most of us. The two freshwater relatives are pulmonate snails. Shells not spiralled but instead form neat little caps. Outer margins are softer than other parts enabling them to grip uneven surfaces firmly. **Lake limpet** and **river limpet** fairly common in ponds and lakes throughout Britain and can often be found on pond weeds.

Operculates

Characterised by hard plate on end of foot which seals animal in its shell when it retracts. Also differ from pulmonates in having gills for respiration. These restrict them to clean, flowing water whereas pulmonates, which breathe aerial oxygen, can tolerate the lower oxygen levels of standing water. Most operculate snails have separate sexes, although Jenkin's spire shell is parthenogenic. Both this species and freshwater winkle actually give birth to live young.

Common bithynia Plentiful in most ponds. Superficially resembles species of *Limnaea* with which it occurs, but it has an operculum which closes off shell when it is removed from water.

Jenkin's spire shell Much smaller. Relatively common in freshwater, having spread inland from estuaries in recent times. In lightly saline waters it plasters the mud at water's edge. Numbers are so astronomical that it has become an important food source for many water birds such as shelduck. Surprisingly it was unknown before 1859 and only first noticed in freshwater in 1893.

Bivalve molluscs

Completely different from freshwater snails. Instead of enclosing body in spindled tube, they secrete two separate hinged valves which can be closed by powerful muscles. Limited powers of movement, possessing muscular foot mentioned previously. Live buried in mud at bottom of ponds and streams. One or two extensible tubes or siphons which allow water current to pass over gills for respiration and feeding. Ciliary action creates current of water moving food particles towards mouth.

Freshwater mussels Several species, most of which are quite large including well-known **swan mussel**. Fairly sedentary, living in mud. This presents problems in long term, since silt settles out of water and would eventually cover them. Thus it is necessary for them to adjust their position from time to time with the aid of a muscular 'foot'. This finger-shaped object can be forced into mud around shell and used to lever animal into a better position.

Separate sexes, the filter-feeding currents drawing in sperm from adjacent mussels. Fertilisation occurs within female and resulting glochidia larvae are ejected in spring. Look rather like miniature adults and can swim around by clapping two valves together rather like scallops. At this stage they are looking for a fish to bind on to. Successful ones initially attach to fish's skin by sticky byssal threads and then proceed to burrow into its skin. Cause obvious nodules on fish and are nourished by its blood. Leave after about three months and settle down to become adult mussels in sediment.

Orb shells and **pea shells** Both much smaller, rarely exceeding 2 cm in length. Orb shells have rounded appearance and two siphons; pea shells have flatter, more uneven shape and only one siphon. Both abundant in silt at bottom of ponds.

Fish

No book on pond life would be complete without a section on fish although they are the animals least likely to be encountered by pond hunters as they prefer deeper waters and quickly escape the clumsy movements of most observers.

Their design is supreme amongst aquatic vertebrates and streamlining enables them to move rapidly through the water. They breathe through gills so they have no need for aerial oxygen. The tail is the main means of propulsion and by movements from side to side, the fish moves forward through the water. A rigid backbone prevents the fish from floundering around and fins, which are strengthened by bone, direct movement. (Dorsal, pectoral and pelvic fins control roll, pitch and yaw respectively.) The swim bladder is filled with air, volume of which can be controlled by absorption or release of gases from blood vessels. This enables fish to remain at any height in the water column.

Gas exchange is performed by five pairs of gills situated behind head. These partly enclosed and protected by plate or operculum,

---

Figure 55 The water scorpion relies on its leaf-like camouflage to allow prey to approach close enough to be caught.

which can either lie flat against body or open. Flow of water over gills and out through opening under operculum is maintained by fish gulping water and depressing bottom of mouth. Gills well supplied with blood containing haemoglobin and gas exchange similar to that occurring in human lungs. Oxygen also absorbed via skin as with frogs; importance of this varies according to species.

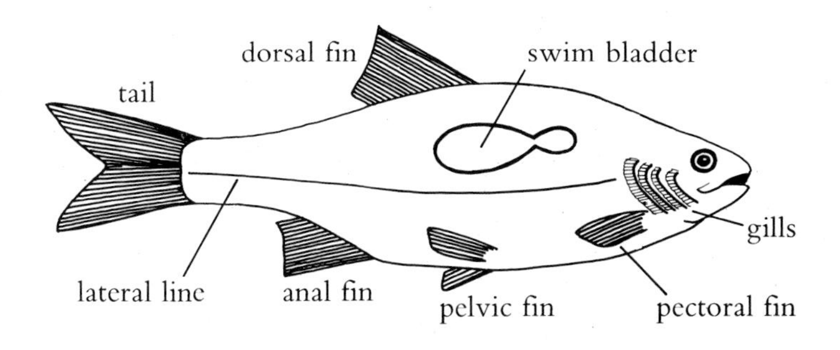

Figure 57 A cross-section through a fish. The fins and tail control movement and the swim bladder and gills enable the fish to remain in the water column and cope with gas exchange respectively.

Carp family or **cyprinids** are well represented in our waters. Include roach, rudd, tench, gudgeon and carp.

Carp Introduced from Europe extensively in medieval times into stewponds where they were grown for food. Can grow to enormous size over several years. Feed by sifting through mud at bottom of ponds and lakes; lips richly endowed with sense organs for taste and touch.

Tench Also essentially a detritus feeder. A rich brown colour with red, beady eyes. Extremely slimy; were called doctor fish, since they were thought to cure other fish by rubbing against them. This may be more than an old wives' tale because slime or mucus which covers fish contains antibodies and plays an important part in their defence against disease.

Roach and **rudd** Shoaling fish. Can be recognised by their silvery appearance and reddish fins. Shoals are particularly noticeable during breeding season when they move into shallow water to spawn. Large numbers of fish can sometimes be seen breaking the water and rolling.

Figure 56 Great diving beetles are voracious carnivores and often tackle prey as large as this three-spined stickleback.

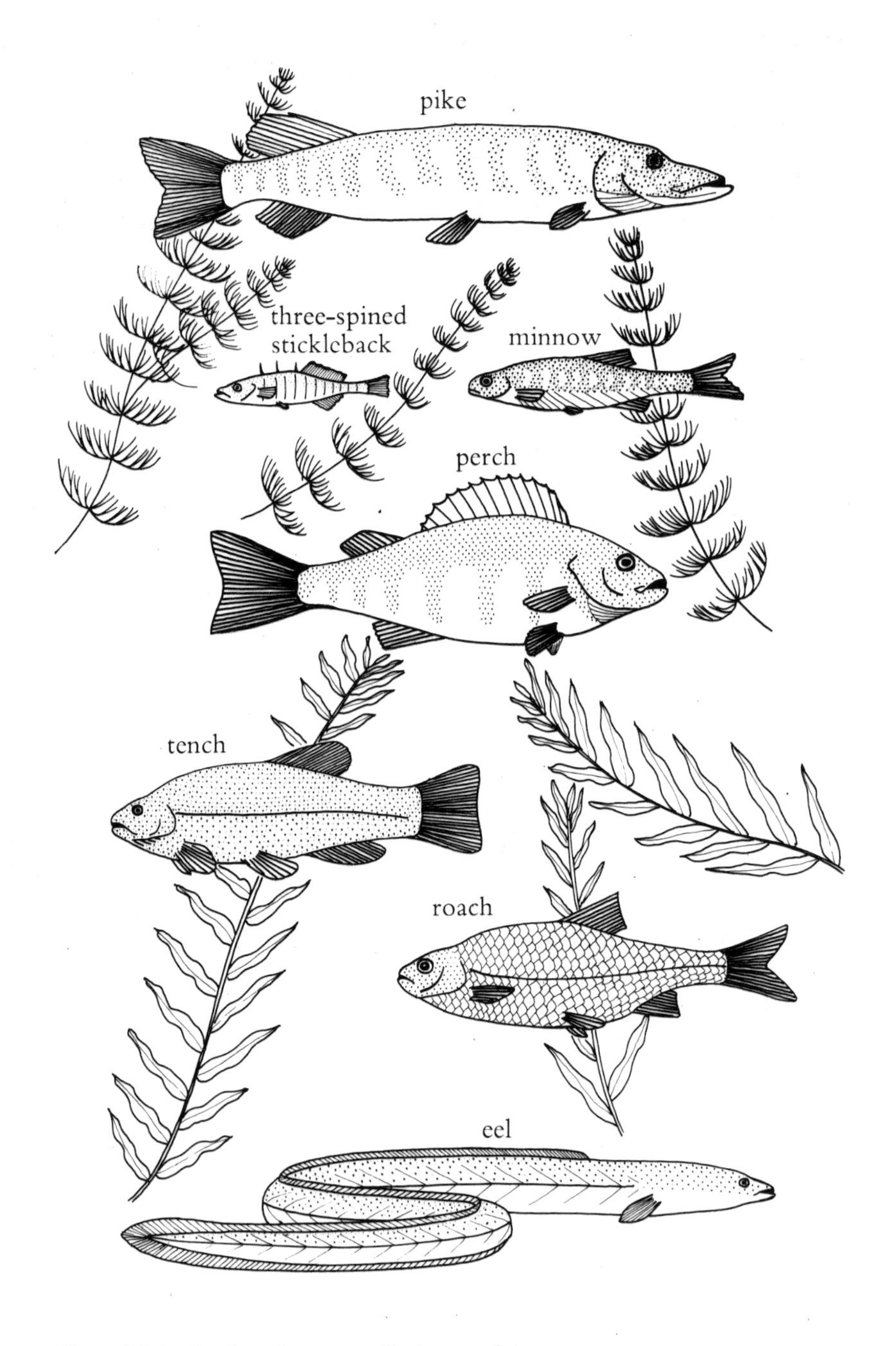

Figure 58 A selection of common freshwater fish.

Perch Another shoaling fish while young. Larger fish live alone and are predatory, even taking small individuals of their own kind. Extremely well marked with lovely green vertical stripes along body and red fins. Stripes provide excellent camouflage amongst water weeds of lakes and rivers in which they are found.

Pike A far more ferocious predator which has earned name of 'freshwater shark'. In larger lakes it is not uncommon for pike to grow to more than 9 kg. At this size they are capable of feeding on fish as large as roach or rudd. Hunting success partly due to superb camouflage. Like perch, pike has vertical stripes on body and uses these to best advantage by lurking amongst weeds at edge of lakes. Body designed for short, rapid acceleration and it will shoot out and grab fish from a shoal. Evidence of this can often be seen above surface as tiny fish leap out of water to escape. Pike has earned a bad reputation amongst owners of lakes especially where these are stocked with coarse fish for angling. They perform a vital role in balance of nature, however, by keeping numbers of other fish in check. In addition, like most predators, pike selects easiest prey, often sick or weak individuals.

Three-spined stickleback Minute in comparison to pike, being about 5 cm in length. Compensates for this by its fascinating courtship and mating behaviour which can be readily observed at home and is discussed in detail in Chapter 6. Three strong spines behind dorsal fin; no scales on body but this is protected by bony plates. Feeds on invertebrates and small fishes. Can be found in most ponds.

Eel Although a true fish, eel seems quite unlike the fish just described. Elongated body with fins much reduced. Scales also much reduced giving it an almost reptilian appearance. Average pond hunter will see very few eels in the course of his studies since they are largely nocturnal and spend day buried in mud or hidden under stones.

For hundreds of years it was a mystery as to how eels bred. Fact is almost stranger than fiction here. They are not spawned in mud at bottom of ponds, nor do they arise from horse-hair, but come from the far distant Sargasso Sea. From here, tiny transparent larvae, leptocephali, are carried by Gulf Stream to our shores. Develop on the way and on reaching freshwater they are small eels. Although they have gills, can also breathe via skin and take up 60 per cent of oxygen requirements in this way which enables them to move across land as long as they are damp. Hence eels may be found in ponds far removed from nearest flowing water. An eel may spend up to ten years in freshwater before returning to the sea to attempt reverse migration.

Lamprey Similar elongated appearance to eels but sucker instead of jaws. Found in stony streams and brooks. Has row of gill slits along each side behind head, instead of single pair found in other fish. Feeds on small invertebrates, eating them whole, but will also parasitise fish. Sucker used to cling on to fish while rasping teeth cause a flow of blood. Evidence of a lamprey attack on fish, a circular scar 1 to 2 cm in diameter, is sometimes visible.

Frogs, toads and newts (Amphibia)

Amphibians are represented in Britain by three species of frog, two species of toad and three species of newt. Are at home both in water and on land but only toads stray far from water, others remaining in damp places throughout the year.

All our amphibians return to water to reproduce and this is when they are most easily visible. Eggs laid in protective jelly and masses of frogspawn a familiar sight to many. Toads produce a string of eggs and newts carefully lay their eggs singly. There may be a thousand or more eggs in a mass of frogspawn, a seemingly huge volume to have come from one animal. As eggs are laid they are coated by a secretion from the frog's vitelline gland. On contact with water this swells to form a jelly-like substance known as frogspawn. Not only does the jelly give the eggs buoyancy and keep them in warmer surface water, it also cushions them from physical damage and deters other animals from eating them. A full description of tadpole development is given in Chapter 6.

In spring male frogs are distinguishable from females, not only by their leaner appearance but also by nuptial pads on thumbs. These suckers enable them to grasp female's slippery body during mating. Mating position in amphibians known as amplexus. Male climbs on female's back and grips her shoulders so that their cloacae are close together. As female lays her eggs, male sheds sperm ensuring fertilisation. Mating in toads occurs in a similar position and large numbers will often congregate in suitable areas.

Newts have elegant courtship display and acquire breeding colours. This described in more detail in Chapter 6. After display, male releases envelope of sperm which female collects with her vent and sperm are shed into her body to fertilise eggs. In contrast to frogs and toads, female newt lays her eggs singly and wraps them carefully in a leaf for protection. They hatch into tadpoles which take longer to develop into adults, often not maturing until following season.

Gills well developed in tadpoles of frogs, toads and newts. Most important in young tadpole and have a good blood supply but as tadpoles grow and mature, gills gradually atrophy. Adults breathe air – amphibians have well developed lungs – but gas absorption via skin also important. Latter method explains affinity of frogs and newts for water or at least wet places. Apart from breeding, water essential for gas exchange to take place via skin which has to be wet.

Toads more adapted to drier habitats and resistant to desiccation.

Tadpoles of frogs and toads and adult newts swim by side to side movements of tail, but end result is not as efficient as in fish, due to lack of stiffening of bone. Adult toads and frogs have webbed feet to increase surface area and with these can propel themselves quite rapidly through water.

Frogs

In addition to **common frog**, two other species have been introduced to Britain from continent.

Edible frog Locally common in some Kent marshes but has not spread far from site of introduction.

Marsh frog In contrast this has spread along south coast marshes as far as Sussex. Unlike common frog, it is found in or around water all year round and even hibernates in mud at bottom of drainage ditches. Has a more pointed head than its common relative and eyes situated more on top of head. This useful since it likes to lie in the water with its eyes just peeping out to watch for predators. Marsh frogs are very sensitive to disturbance and will submerge with a plop at slightest hint of danger. During a warm summer's day distinctive loud croaking can be heard from some distance. They often sit on the banks and sing although also do so from sanctuary of water. Main alarm signal to marsh frogs is movement and if you sit quietly beside a ditch they will re-emerge after some time if you keep still. Movement also provides main impetus for feeding, as with the other amphibians. Not until an animal moves is it recognised as food. Both frogs and newts feed largely on invertebrates.

Toads

Rare **natterjack** toad distinguishable from **common** toad by its smaller size and yellow stripe down its back. It frequents dry, sandy areas such as sand dunes, flocking to pools in breeding season. Often water is at such a premium in these areas that any pool will be absolutely alive with toads.

Newts

Smooth, palmate and **great crested newt** found in Britain. Outside breeding season can be found well away from water, often under stones and fallen branches during the day. Return to water during spring and small, shallow pools can contain hundreds.

5 Freshwater ecology

Ecology is the study of the relationship between an organism and its environment. At its simplest level the main factor linking organisms to each other may appear to be food. Even this relationship is a complex one because, for example, predators not only regulate prey numbers but are themselves limited by prey numbers. If too many prey are consumed, there will not be enough to feed the predators' offspring. Plants would seem to depend solely upon light and nutrients for photosynthesis. However, they could not survive if animals and bacteria did not aid the recycling of nutrients.

Food is only one important factor in ecology. Animals often depend upon plants for shelter from predators and protection for their young. In the freshwater environment, for example, the leaves of water lilies give shade from sunlight and a substrate on which snails and China mark moths can lay their eggs. Plants may also provide camouflage for various animals either to escape predation or to conceal themelves whilst waiting for prey.

The chemistry of the water and the soil type of the surrounding area all affect the organisms which will be found in a pond, as has been discussed in the first two chapters. Finally, both animals and plants have adapted to cope with seasonal variation and with the inevitable natural filling in of the pond.

The food web

All life is dependent upon plants. Without their production of the essential gas oxygen, most forms of life would cease to be. In ecological terms plants are primary producers. All plants from the simplest unicellular alga to large, complex plants (called macrophytes) such as water lilies produce oxygen by photosynthesis. This involves the fixing of light energy by the green pigment chlorophyll (which gives plants their colour) and the conversion of carbon dioxide and water to form glucose and oxygen. As carbon dioxide is an end product of animal and plant respiration and oxygen is a requirement of this process, both respiration and photosynthesis are interdependent.

Many protozoa and microscopic organisms depend upon the

unicellular algae that make up the freshwater plankton. They are termed first order consumers and they may in turn be eaten by other order organisms, for example *Hydra* and mosquito larvae. These second order consumers are active carnivores. As you can imagine, things can become very complicated after this point. Dragonfly nymphs may eat mosquito larvae, fish may eat the nymphs and birds may eat the fish. Relationships between organisms do not simply form a chain but rather a complex web. Any attempt to express this on paper is inevitably a gross oversimplification.

All animals that are permanent residents of freshwater will ultimately die naturally or be eaten there. A great many of the temporary visitors will inadvertently never leave. Together with the waste products of all the inhabitants, this means that there is a great deal of organic material or detritus entering the freshwater system. It is here that the biological 'vacuum cleaners' or detritivores come into play. Many fly larvae and snails perform this useful function and their waste products may in turn be consumed by other detritivores or undergo bacterial breakdown. Bacteria and fungi are the ultimate disposal organisms and many of the detritivores rely on bacterial decay to 'soften up' their food.

Although a certain amount of bacterial and fungal decay is essential to ensure recycling of nutrients, too much can have a devastating effect. If the water is artificially enriched with organic waste such as sewage this encourages, in particular, bacterial division. Since many bacteria respire oxygen their activities can seriously deplete oxygen levels in the water. This can have a harmful and often fatal effect upon the other inhabitants of the water if they cannot satisfy their oxygen demands. If organic fertilisers run off into a freshwater body they can encourage the growth of single-celled algae, a process known as eutrophication. This too can have a detrimental effect on pond life because light penetration is reduced killing off other plants and bacterial action is encouraged. The effects of pollution are more fully discussed in Chapter 7.

The wealth of dead organic material is never more obvious than in late autumn when many of the larger plants die back and the freshwater system can appear almost choked. By the following spring most of this will have been broken down and the nutrients recycled. This recycling is vital for the survival of a pond. It is easy to observe in Britain since most freshwater plants are annual and the animals' life cycles are often adapted to cope with this.

Although many pond organisms can be neatly defined according to whether they are first, second or third order consumers, some are not so easy to categorise. The same species of leech, for example, may be classed as a first or second order consumer as well as a detritivore. Frog tadpoles start life as algal grazers and later on become voracious carnivores.

Other animals have an opportunistic way of life and take whatever is going. Pond skaters can frequently be seen gliding over the surface of water. Their legs are so sensitive to vibrations that they will rush to a butterfly that has just fallen into the water. They are also quite happy to take the remains of a long-dead fly should they chance upon it. Underneath the surface, the water shrimp can be an active carnivore but it will readily take the leftovers from someone else's meal.

Parasitism

A parasite is an organism that feeds on another plant or animal 'host' without necessarily killing it. Most plants and animals are subject to attacks by parasites and their populations are adapted to cope with parasitism just as they are to predation. Parasites are obviously not beneficial to the individual they attack but the inter-relationships are significant and worth considering. To some degree parasites keep a check on animal and plant numbers as do ordinary predators. The fewer hosts around, the fewer parasites survive, and a sudden increase in host numbers invites a similar increase in parasites. As you might expect, a rise in the number of hosts is followed by a delay, before parasite numbers increase. With all the life teeming under the surface of the water, it is hardly surprising then that many parasites are found. 'Worms' are traditionally regarded as the typical parasite by the layman but many other types of animal have also adopted this life style, such as the fish louse which is a crustacean.

Invertebrates often suffer the attacks of parasites. Water bugs such as the water scorpion and the water stick-insect frequently have parasitic mites attached to them, especially at a joint where the cuticle is thinner. It is an extraordinary sight to see a water stick-

Figure 59 Vertebrates often suffer attacks from parasites. This ten-spined stickleback has a parasitic cyst on its flanks behind the pectoral fin.

insect sucking the juices from a water shrimp, while it, in turn, is having its body fluids drained by half a dozen larval mites.

Mosquito larvae often infect areas of stagnant water and the edges of ponds and lakes. The larvae are sometimes parasitisised by a nematode or roundworm. This starts life comparatively small in relation to the mosquito larva, but before long it grows to a large size and can be seen coiled around inside the thorax of the larva. The parasite can be up to a third of the size of the larva and will eventually kill it. It is not surprising that this sort of worm has been considered as a form of 'biological control' for mosquitoes in countries where they transmit diseases such as malaria. In this instance the parasite is serving a useful function to Man by killing off hosts which transmit disease.

As well as having a parasitic life style, many nematode worms can be found free living in detritus on the bottom of the pond. The flatworms or platyhelminths, in contrast, are almost exclusively parasitic. Many tapeworms and flukes are found as adults in fish or amphibia but often they will have more than one stage in the life cycle. The larval stages are frequently found in animals that are food items for the host of the adult which aids the transmission of the parasite from one host to another.

An example of a complex life cycle is that of the cestode *Ligula intestinalis*. The adult tapeworm lives in the gut of fish-eating birds such as the great crested grebe. The adult parasite only lives for a short time, but produces thousands of eggs which upon hatching, infect copepods and develop into a larva called the procercoid. Eventually an infected copepod such as *Cyclops* spp. will be eaten by a fish such as the roach. The larval parasite then leaves the copepod and burrows through the gut wall of the fish and grows into the plerocercoid larva in the body cavity of the fish. This parasite is particularly interesting because it increase its chances of being eaten by the right host in two ways. It gives off chemicals that 'parasitically castrate' the host fish. This reduces its tendency to shoal, particularly during the breeding season, thus making it an obvious target for a bird. Also, if the parasite grows big enough (they can weigh over 20 g!) or there are many of them, they impair the operation of the swim bladder of the fish, causing the fish to swim on its side on the surface of the water – a sitting target. When a bird eventually eats the fish, the life cycle is complete and a new adult worm develops.

Fish are also prone to attacks by parasitic fungi. You may have noticed woolly growths on goldfish in your garden pond. This is a parasite called *Saprolegnia* and it attacks the fish only after it has been damaged. The slime which covers a fish's body is an important part of its defence and contains antibodies to combat infection. If some of the mucus or a scale is removed then this leaves a path for the fungus

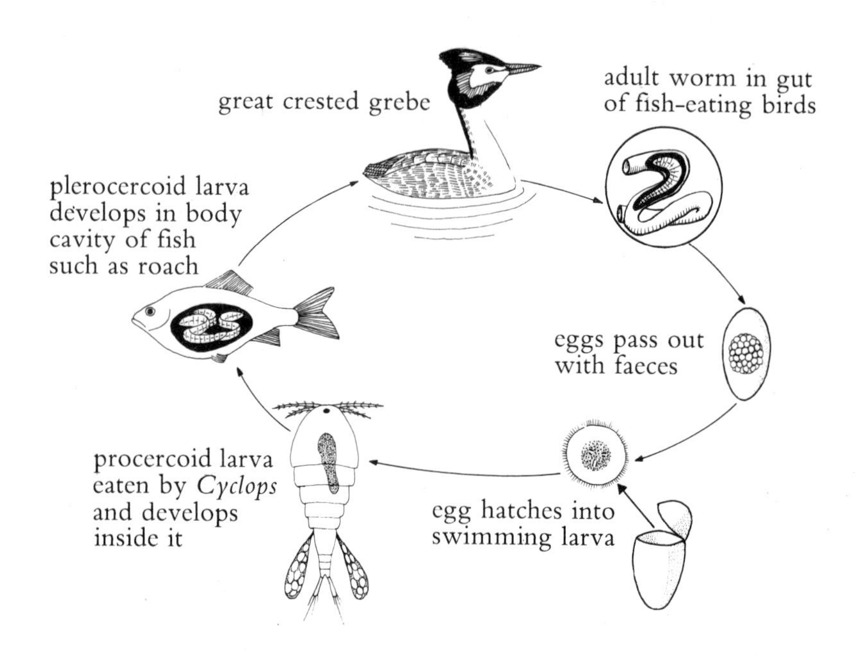

Figure 60 The life cycle of the fish parasite *Ligula intestinalis*.

to enter the fish. Unfortunately fish that are put back by anglers are all too often damaged in this way.

Simply removing parasites from any environment would then have a harmful overall effect because host numbers would increase dramatically and, in turn, detrimentally affect their own food or prey. As such, parasites should be regarded as specialised forms of predators which play an important role in maintaining the ecological balance of populations.

Predator / prey relationships

The predators of the aquatic ecosystem are inevitably dependent upon the continued existence of their prey – the fewer prey, the fewer predators, in simple terms. This is called a predator-prey relationship. The succession of orders of producers and consumers is linked to a 'pyramid of numbers'. At the base is a vast abundance of producers. Fewer numbers of first order consumers are present to use this resource. As you move up the pyramid, less and less individuals are able to take advantage of the preceding level in the pyramid. If there were a similar number or more predators of an animal than prey, then the prey would soon be depleted to a point where the predator could not survive. Although numbers of predator and prey fluctuate from year to year, the overall balance remains the same.

Each organism in the pyramid fits into what is called a 'niche'. This defines its position in the food web in relation to organisms that it consumes and is consumed by. Many animals may appear at

Figure 61 The mouth of the lamprey forms a sucker with which it can suck blood from other fish such as this trout.

first to occupy the same, or very similar niches. For example, beetle larvae and dragonfly nymphs are both carnivores and might appear to compete for the same food. However their approaches to catching prey are different. In general, dragonfly nymphs such as that of the emperor are stalkers and rely largely on stealth amongst floating vegetation. Beetle larvae are continually active, seeking out prey as large and mobile as small fish and tadpoles and they generally feed nearer the bottom.

Dragonfly nymphs have evolved a strange, but very effective way of catching prey some distance away from them. Underneath their heads the cuticle forms a long extension which is folded back on itself and lies pressed against the head. This 'mask' as it is called, is hinged at the fold and where it joins the head and it can be rapidly shot out, to a distance of a centimetre or more. On the end of the mask are two teeth which pierce and hold the prey until the mask retracts and the food can be attacked by the mouthparts. The nymphs are masters in the art of stalking and with stealth they can approach unsuspecting tadpoles or fish, or even other nymphs, close enough for the mask to be effective. The most developed sense of the nymph is vision. The compound eyes take up a large proportion of the total head area and in the emperor dragonfly, appear to wrap around it. These eyes give them an acute idea of any movement around them. If you wave a pencil or your finger in front of a hungry dragonfly nymph, you will notice almost imperceptible movements as it orients itself toward the object. Slowly, it will move closer and closer, until eventually it will flick out its mask, often again and again, until it finally decides that what it is attacking is inedible.

The various species of dragonfly nymph might be expected to

Figure 62 Great diving beetle larvae are active carnivores and this one is devouring a tadpole.

compete with each other but if studied carefully, you will find that they have different niches. Nymphs of the broad-bodied libellula are broad and hairy and half bury themselves in the mud and leaf-litter at the bottom of the pond. They wait here for passing prey, which they catch with the aid of their mask. At the other extreme, nymphs of the emperor dragonfly stalk more actively amongst floating plants near the surface and so these two species rarely encounter each other. Other species live at different levels in the pond and so only overlap to a small degree. This means that they do not compete for the same food, or at least not in the same location.

Figure 63 The nymph of this broad-bodied libellula dragonfly is well camouflaged in pond mud as it waits for prey to pass by. Most of its body is hairy and collects silt, but the eyes can be seen clearly on the left.

Zonation

Zonation is most obvious on the seashore where the tide produces a marked gradation in the animals and plants according to their tolerance of exposure to air, seawater and other factors. As you can imagine, the zonation in a pond is not as clearly defined, or at least, it is not as easy to observe. However, just as animals can be categorised according to how they feed, they can also be zoned. This helps *us* define their ecological niche, and helps *them* avoid competition for the same food and places to reproduce. Plants can be zoned in a similar way.

As freshwater is not tidal, the water level remains the same for longer periods of time than in the sea and plants can grow in a more stable environment according to their tolerance of air and water. Light is another important factor governing plant zonation since it is absorbed by water. Thus in the water the plants are adapted to photosynthesise using less light; surface loving plants would not survive as well in this position. Temperature is also governed to a certain degree by depth and plants found in deeper water will be less tolerant of changes because temperature variation is lower here.

The surface of the pond is the vital interface between water and air where gas exchange occurs and there is the most light. In ecological terms it is known as an ecotone. The surface film has its own special plants and animals, many of which take advantage of the surface tension which allows them to float. Pond skaters are the supreme example of animals adapted to life on the surface, catching prey trapped in the water. Their sensitivity to vibrations not only allows them to detect struggling prey but also indicates danger. If you vibrate a grass stem gently in water, pond skaters will come rushing to it, but if violent movements are made they will flee. Many animals, such as the water boatman, use the surface zone of the water but prefer the sanctuary of the under surface. The water boatman detects the movements of stranded insects but will take its catch down below to enjoy the meal alone.

At the air/water interface the diffusion of oxygen into the pond and carbon dioxide out occurs. Some animals, however, still use oxygen from the air and must return to the surface from time to time to replenish their air supply. Some, such as the rat-tailed maggot, have breathing tubes that are simply thrust through the surface, but

Figure 64 This pond skater is feeding on a fly trapped on the surface film. Skaters have a waxy cuticle and rely on surface tension to keep them afloat.

others, such as mosquito larvae, have an intricate arrangement of hairs around the tip of the abdomen. These fan out when the abdomen touches the surface, using the surface tension to allow a continual link between air and the abdomen. Living at the surface makes them potentially prone to predation from both aquatic and terrestrial animals. However, both mosquito larvae and pupae are very sensitive to disturbance and can relinquish their hold on the surface and swim downwards with wriggling movements of their abdomen.

The next zone down in the pond is that of the floating vegetation such as Canadian pondweed. For many animals, such as damselfly nymphs, these plants provide protection and shelter. They are often well camouflaged and thus escape predation. Other animals use the camouflage of the vegetation for predatory purposes. The water scorpion is shaped like a leaf, the air tube even resembling a stalk. It lies motionless amongst the weeds until some unsuspecting animal comes within range. The water stick-insect is also camouflaged resembling a stick. Unlike its terrestrial counterpart, which uses its camouflage to escape predation, the water stick-insect awaits its victims in this way.

Beneath the floating weeds is the open water. This is the kingdom of the fish in particular, but of course many other organisms pass through it on their way to and from the surface. In a lake the open water will make up a significant part of the total volume of water, but in a pond according to definition, this zone will be of approximately the same depth as that of the rooted vegetation. This region provides shelter and protection for animals. To many, the plants provide a substrate over which to move. Snails in particular use the plants for getting to the surface to replenish their air supply and many insect larvae found in ponds like to clamber through this vegetation and are adapted to conceal themselves here. The nymph of the hairy dragonfly has a flattened, almost concave ventral surface and a smooth dorsal surface enabling it to cling tightly to reed stems; at a casual glance it looks just like a bulge in the stem.

Sinking lower in the pond we come to the surface of the mud. It is here that decaying plant material falls along with dead and dying animals and this is the haunt of the detritivores and scavengers. Myriads of ostracods, the minute pea-shaped, bivalve crustaceans, scurry over the surface, feeding and laying eggs on decaying leaves. Larger animals such as the water louse rummage through the detritus together with leeches and many species of snails, seeking items of any food value.

Beneath the surface of fresh detritus is the bottom of the pond which is composed of mud and advanced decaying organic remains. The mud provides a firm substrate and is a relatively safe place from predators. The anaerobic nature of bacterial decay and the density of

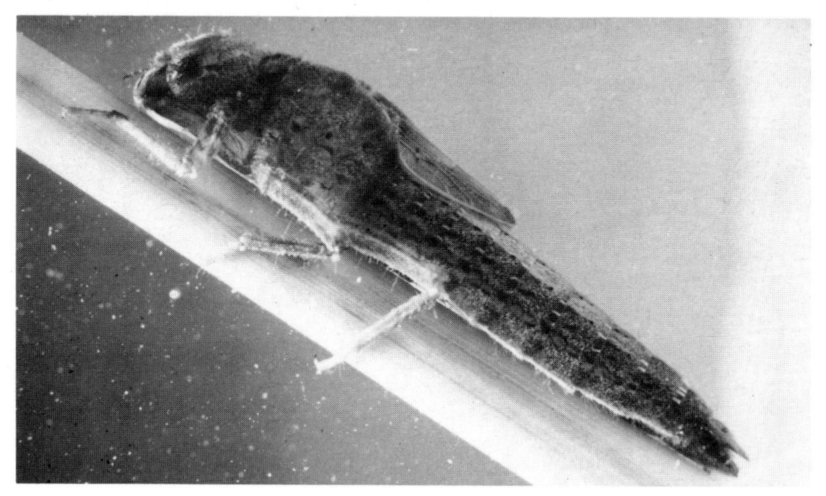

Figure 65 The nymph of this hairy dragonfly is well camouflaged. When it presses its body close to a reed it looks superficially just like a bulge in the stem.

the mud makes the pond bottom low in oxygen and many animals have therefore evolved to obtain oxygen from the water above. The pea mussel siphons water from the water above using long and mobile tubes which channel the water over its gills. Other animals such as chironomid worms have respiratory pigments like haemoglobin which can trap oxygen when the levels are low. All the animals in this zone have to cope with low oxygen levels, but living here gives them an advantage as well. Many other pond dwellers are totally unable to survive at the pond bottom and so those that do will not be competing with them.

Seasonal variation

Anyone who has dipped in his local pond may have noticed that, as in all habitats, the numbers of various animals and plants vary throughout the year. Britain experiences extremes in conditions from summer to winter in terms of temperature and light. Aquatic ecosystems reflect the seasons and a number of different adaptations have evolved to cope with the hazards of winter and take advantage of the abundance of food and light in the summer months.

Some pond dwellers, such as amphibians, inhabit ponds only during the summer months. However amongst the frogs, toads and newts are found varying affinities for water. All of them must return to water for breeding, but in the case of toads, they leave soon after mating has occurred. The common frog and the three species

Figure 66 The endearing water spider is common in many ponds. The silvery appearance of the abdomen is due to a coating of air trapped by hairs which helps it breathe underwater.

of newt can be found close to water for most of the year but will leave it for considerable periods of time. The marsh frog has a greater affinity for water than its common relative and it even overwinters in the mud at the bottom.

Many pond dwellers are obliged to spend their whole lives in the water because they are unable to survive on land. The protozoa, snails and crustacea all live and reproduce in the pond.

Figure 68 Frogs often collect in vast numbers in suitable ponds for mating and egg-laying. The two in the foreground are in the mating position known as amplexus.

Reproduction takes place during the summer months on a more or less continual basis, eggs being produced in large numbers. Many snails protect their eggs in cases to prevent them being eaten or damaged. Some crustaceans, such as copepods, go one step further and carry their eggs around with them until they are ready to hatch, giving them an even greater degree of protection. During the winter months, all these creatures will be much less active and feed less often.

Although freshwater insects are secondarily adapted to water, most of them have evolved to spend the winter in the pond and do not seek the sanctuary of land. Temperatures can be extremely low in the winter and food may be scarce, so in those insects which have a one-year life cycle (univoltine) the eggs overwinter, remaining dormant until spring. Large species of insect may take several years for the larvae or nymphs to become fully grown because they cannot grow as quickly in winter as in summer and their life cycle is termed multivoltine. Some species of dragonfly nymph become

Figure 67 Swamp spiders are a magnificent sight around heathland pools. They detect vibrations of drowning insects through their legs, and this female has caught a damselfly.

relatively inactive in the winter and feed infrequently, whilst other insects may even hibernate in the mud.

Dispersal

As discussed earlier, a pond is a doomed habitat, destined to fill in eventually. Depending on its size this may happen sooner or later, but all the inhabitants must come to terms with it. Therefore it is essential that pond organisms are continually dispersing and colonising new habitats in case the old one dries up. Dispersal is not so important for animals living in large water bodies such as lakes and gravel pits which remain stable. Colonising a new area also avoids competition in overcrowded waters. The ability to move through water does not solve the problem of dispersing to new habitats which requires different adaptations.

Amphibians overcome this problem by being able to leave the water at will and survive quite adequately without it. They must, however, return to breed, and often have remarkable homing abilities, notably the natterjack toad. Some individuals will find new ponds and colonise fresh areas and because of their relative independence of water, frogs will often try to breed in very temporary areas.

An efficient means of dispersal is by air, and here insects are the masters. Many do not have to fly except to disperse. Water beetles and bugs are true pond animals but in summer months they may fly some distance from water. The great diving beetle will sometimes land on shiny surfaces such as car roofs and greenhouses mistaking these for areas of water.

Many insects, however, are obliged to spend their adult lives out of water, and this is their dispersal stage. In mayflies the adult life is very short – a matter of a day, during which they do not feed but simply mate, lay eggs and die. Dragonflies, however, may spend many weeks flying above the water surface. Often, when a dragonfly emerges from its nymphal skin, its maiden flight takes it well away from water, where it may spend some considerable time. It will then return to water, which may or may not be its original pond. Here it will have to contend with other dragonflies, for they are extremely territorial.

If you sit beside your local pond on a calm summer's day, you may see and hear the loud clatter of dragonflies' wings as they do battle for their particular patch. Species such as the emperor and the broad-bodied libellula are particularly territorial and will not let members of their own or other species into their area. By preventing others from entering their territory they ensure that only their offspring live in that part of the pond and that they will not be competing for food with other dragonflies. These battles may even result in the death of an individual and on several

Figure 69 A mayfly hatch on a warm spring evening. The males swarm together in vast numbers; they only live for a short time.

occasions I have seen an emperor dispatch a sympetrum.

Many of the crustacea and simple pond organisms have resistant stages in their life cycles which are often eggs. These can survive both desiccation and freezing. In two primitive crustaceans, *Triops cancriformis* and the fairy shrimp, the eggs will not hatch until they have endured drought and cold. This is because they live and breed in the most temporary pools and hatch only when the pools have refilled with water and before they have dried up in summer. This

Figure 70 Territorial disputes between dragonflies are often violent and can result in wing damage as has happened to this four-spotted libellula.

Figure 71 Fairy shrimps colonise temporary pools and vast numbers may be produced in the summer before the water dries up. Their eggs are resistant to desiccation and can be blown to new ponds.

also means that the eggs can survive being blown around and carried by birds and hence colonise new areas.

But what of the fish? Many species have eggs that will survive for a certain period of time out of water and it has been suggested that they could be carried by water birds from one pond to another. Indeed it is often a mystery as to how a new pond can have fish in it when it has not been deliberately stocked. Fish eggs will undoubtedly be transferred with pondweeds if these are introduced, but some may be carried by birds into new areas. Methods of examining the rate at which colonisation occurs in a new freshwater area are discussed in Chapter 6.

6 Pond and aquarium projects

The practical side of pond dipping can provide immense pleasure to people of any age and many happy hours can be spent merely examining the contents of the net. However, if you have a specific aim when you visit your pond, this can be even more rewarding. By carrying out particular projects you can gain a much greater insight into the variety of forms and behaviour of the many pond organisms. In this chapter I hope to give some guidelines to help you to get the most out of your own studies and also to suggest some feasible projects which you can carry out.

The pond itself is an ideal place to learn about the distribution, abundance and behaviour of pond organisms and an aquarium at home also enables you to watch many animals behaving naturally at your leisure. More specialised ways of studying pond organisms, such as photography and the use of the microscope, can give further pleasure and different insights into pond life. At whichever level you choose to study your pond, the recording of information increases your enjoyment tremendously.

It is a good idea to keep a notebook in which to log the events of the day. This should contain details of the date, time of day, approximate air temperature and the overall weather conditions (for example if it is sunny or raining). These facts may correlate with the abundance and distribution of certain animals, as will be discussed later. If at all possible, take identification books to the pond so that you can identify your catch on the spot. (See Further reading for suggestions.) Do not take pond organisms home unless you are sure that you can look after them properly and they will not be left to dry up on a windowsill. If you do not have any books with you, then the next best thing is to put the animal or plant into a white tray and sketch it. In the case of an animal, you could also note how active it is, how it swims and most importantly, its size, as all these points will help with later identification. Size can be very deceptive, especially if you are trying to remember something seen several hours previously in the field. Another helpful pointer to the identity of plants and animals is their distribution within the pond. For example, was the organism found at the surface, as would be the case with duckweed or a water boatman, or on the bottom, as with a pea mussel? All these factors should help in identification and your notes can form the basis for learning in more detail about the distribution, abundance and behaviour of pond organisms.

The distribution of pond organisms

Mapping plant cover

It can be very rewarding to draw a simple sketch map showing the distribution of the floating and emergent plant species in your pond as this can be related subsequently to the distribution of the animals. Certain plants growing around the edge may form distinct zones and the detailed analysis of zonation is discussed later.

You can mark on your map any trees around the edge of the pond and where the bank is shallow or steeply shelving. If you can identify the plants, then mark in any large areas of single species. You will probably notice that where you find one plant species, such as frogbit, it will be quite abundant in that area. Conversely, in another area you may find no specimens of that species but lots of another. These are known in ecological terms as 'pure stands'. It is often the case that the position of these stands can be correlated with areas of shade or depth of water. Many animals prefer to live in association with one particular plant species. By estimating the abundance of the various animals in relation to plant species you can discover these preferences. The various methods of estimating abundance are dealt with in a subsequent section.

If you would like to produce a more accurate map of your pond, this can be done relatively easily. You will need two lengths of string which exceed the maximum width of the water. Tie a piece of coloured twine at 1 m intervals along the lengths. These markers will be easily visible from a distance and can be secured by tying a knot around them in the string. You then have a pair of home-made tape measures.

Begin by stretching one length of string along the longest side of the pond if it has one. Secure this by tying each end to a stake in the ground. Next tie the ends of the other length of string to stakes. Place one stake in the ground alongside the first piece of marker twine on the fixed length of string and stretch the free end across the pond, staking it into the ground so that the string is approximately at right angles to the first one.

You can now start drawing a map using either graph or plain paper on which you have marked a grid. The scale is important: depending on the size of the pond, something like 1 cm on the map can represent 1 m of water.

Starting at one end of the second piece of string, work your way along its length marking on your map any significant plants, or the edge of the pond at metre intervals. For example an area of starwort

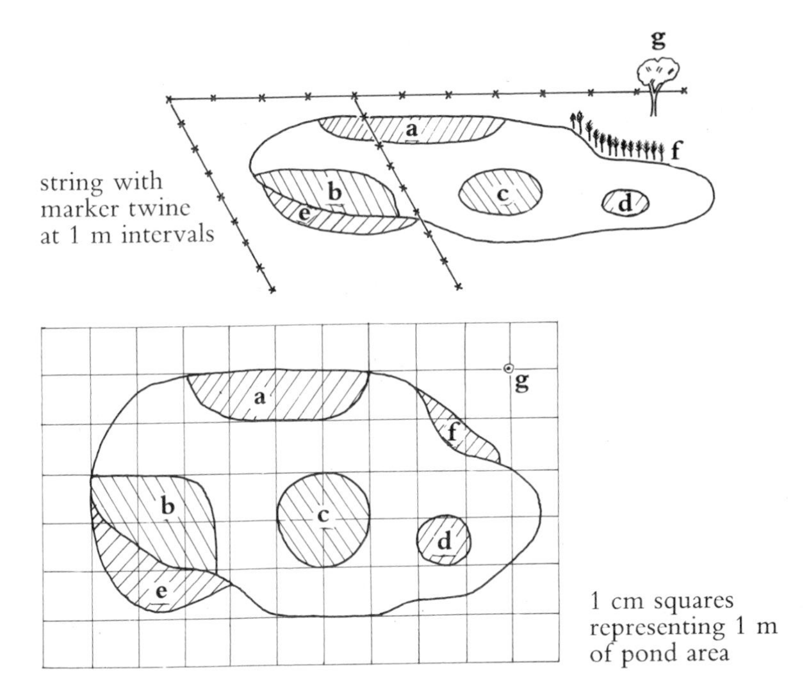

Figure 72 You can map the plant cover on the pond surface using home-made tape measures and marker twine. The diagram below shows an example of a finished map, with each of its centimetre squares representing one metre of the pond surface. The shaded areas depict the areas occupied by each plant species.

between the 3 and 5 m marks on the string would be marked on the map between the third and fifth centimetre. When you have finished recording the plants from along the length of string, move this second length along a metre at both ends. The various features of the pond and its vegetation can be marked off again. When you have moved along the whole length of the pond in this way, you will end up with a complete map of the pond and its vegetation.

This method is obviously impracticable if the pond is large, or if the terrain prevents you from getting around the whole circumference. If this is the case, and you would still like to do an accurate vegetation survey, then you could make a transect.

Making a transect

A transect is essentially similar to a single line of the grid map. However, whereas the map looks only at the surface vegetation, a transect is a study of the distribution of submerged plants also. Take a length of string about 5 to 10 m long with markers tied at ½-m intervals. Stake one end into the ground amongst the vegetation on dry land, attach the other to a piece of cane and plant this in the bed

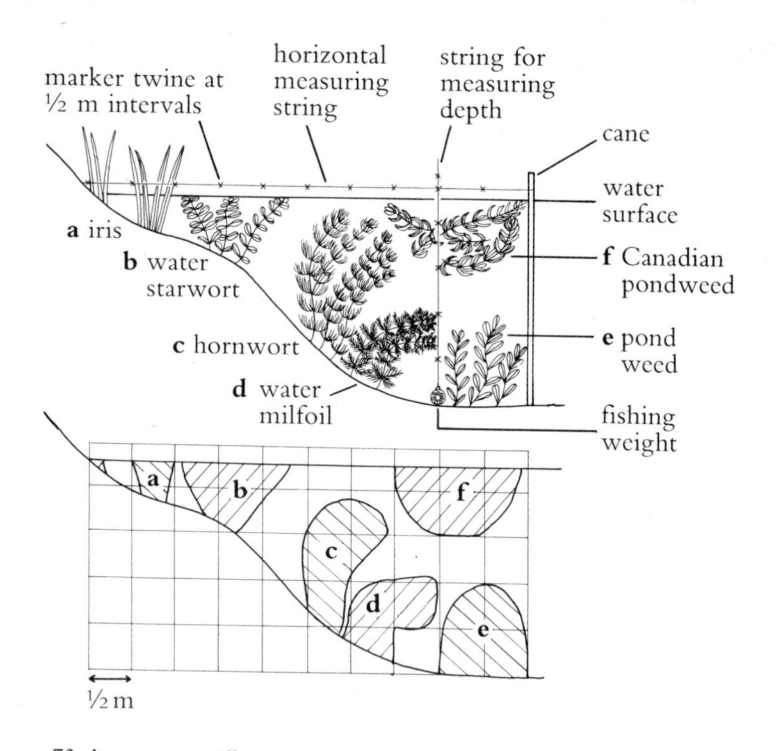

Figure 73 A transect will enable you to measure and illustrate plant zonation through the pond margin rather than just on the surface. The drawing above shows a cross-section through the pond and the position of the measuring string and cane. Below, the finished map shows the areas of plant species and the grid system is drawn to a scale of ½ cm representing ½ m in real life.

of the pond some distance from the edge with the string running along the surface of the water. Obviously your ability to map all the zones will depend upon the depth of the pond and how quickly it shelves. (Take care as you wade from the bank as the bottom may be uneven.) If you are lucky, you should be able to cover all the major zones of the pond's edge before it gets too deep.

On paper, mark a suitable scale, for example 2 cm representing ½ m on the transect string. As you are covering a smaller area than when plotting emergent vegetation, you should use a larger scale on your map. At each marker point, note down the plant species present. You will then be able to see how each plant is distributed and the overlapping of the different species. By positioning the transect string along the water surface, you can also record the depth of water in which the plants are growing. Tie a heavy fishing weight to the end of a piece of string and at each marker point along the transect, lower the weight until you can feel it touch the bottom of the pond. Hold the string at the point where it meets the water surface, take it out and measure the length immersed in the water. You can record this on your transect diagram using the same scale. The same method can also be used to measure the depth of floating plants and the height of rooted plants. You should end up with a useful profile of the pond margin and an accurate guide to the plant distribution.

Zonation

Just as the plants around the pond's edge can be readily grouped according to their distribution, so pond animals can be zoned according to where they are found. You should be able to discover some correlation between the plant zones that you have found with your transects, and the animal distribution. As discussed in Chapter 5, many animals prefer to live in association with one particular plant species and in addition to this distribution, there will be a vertical zonation in the water itself.

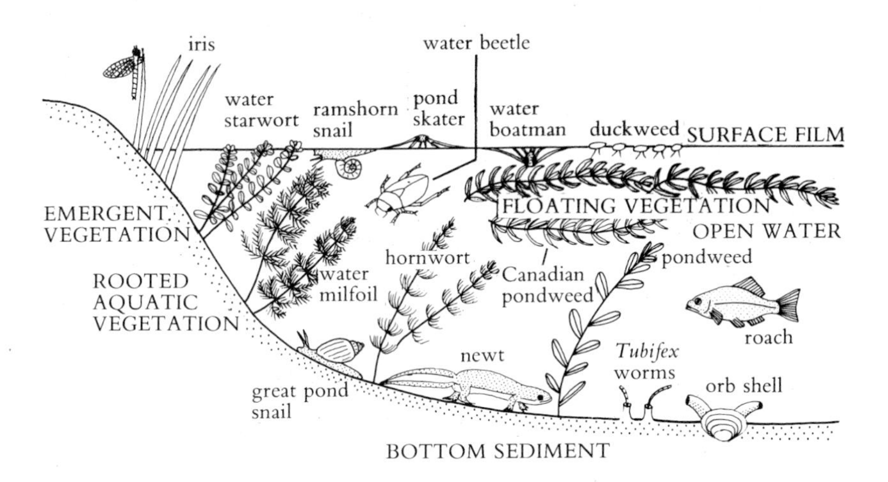

Figure 74 This cross-section through a pond margin shows the horizontal and vertical zonation and some representative plants and animals for each zone.

There are five main vertical zones which you can study. The surface film is particularly important as an interface between air and water. As discussed earlier, many plants and animals exploit the surface tension and many species of water bug and beetle will be found there. You may have some difficulty in separating the animals on the surface film from those in the water immediately below. Since you will probably be using a net to sample, try not to immerse it too deep to avoid confusion.

Below the surface layer is floating weed. It is important to note the plant species, because of the preferences animals often have for particular types of vegetation. If the pond is deep enough, there may be a zone of open water which will contain the more active swimmers such as the great diving beetle.

The last two zones are on the pond bottom. Some animals live on the surface of the mud, while others live buried in it. Again, you may have difficulty in distinguishing between the two areas. The larger animals on the surface can be observed *in situ*; the smaller ones can be pipetted off. The pipette should have a wide bore and an eye dropper is quite suitable. It is best to suck in water rapidly because many pond organisms, even small ones, can often move quickly if alarmed. If you use a net carefully, you can run it over the surface of the bottom without picking up too much sediment. Those creatures living in the mud can be sampled by sieving.

There are two basic ways of recording this vertical zonation in your notes. The simplest method is to score the presence or absence of species. Put a list of species down the left hand side of the paper and the zones examined across the top. You can then mark the presence or absence of a species in the form of a plus or a minus sign. Alternatively, you could draw a line across the zones where the animals or plants were found. The result will look like a disjointed ladder. Be sure to note the sampling method used in each zone as this will enable you to know whether you can make direct comparisons between them.

Time of day

The time of day is another important factor influencing the distribution of certain animals and it should always be noted down in your records. For example, as the sun heats the waters of a shallow pond, there will be a marked increase in temperature and a decrease in oxygen content. Thus, animals that rely on, or supplement their intake with aerial oxygen, such as water beetles, may be found at the surface more often during the middle of the day when it is hottest. You could try making three visits to your pond: at dawn, during the middle of the day and at dusk. At dawn you might observe many newly emerged dragonflies taking their maiden flight, or young froglets leaving the sanctuary of their pond in search of terrestrial insects. In the middle of the day, water bugs and beetles will be most active, coming to the surface for air. Some water bugs may even leave the water and take to the air. Dragonflies will also be making frequent forays to catch insects on the wing and defending their territories. Below the surface of the water the water snails can be observed replenishing their air supplies, but much of their time may be spent sheltering in the shade of plants such as water lilies. In the evening, adult dragonflies may be seen coming to roost on their favourite posts whilst their nymphs are making exploratory excursions up twigs in search of suitable emergence sites. This is also the best time of day to observe mayfly hatches along the banks of chalk streams.

Estimating animal abundance in your pond

If you can combine an estimate of animal abundance with estimates of plant and animal distribution, you will begin to build up a picture of pond ecology. Relating these two factors will give you a clear insight into any plant-animal associations and it could also tell you about habitat preferences. Also if you found that the distribution of a certain predator correlated well with another species of animal, then you could deduce that this might be its favourite prey. For example, where you find the nymph of the broad-bodied libellula, you will also notice the freshwater louse in abundance.

The simplest method is to create an arbitrary scale and estimate the animals present from the numbers in an average sample. The scale could be as follows: A-D with A = 1-5 individuals, B = 6-10, C = 11-20 and D = greater than 20. It is important only to compare samples that have been collected by the same method. This approach is suitable for netting or sieving but you must be careful to standardise your technique. For example, if netting, try to sweep the same length of water, say 1 m every time. If using a sieve, collect similar volumes of mud from the different sites.

The second way of estimating abundance is to use a mark-recapture technique. The principle of this method is first to catch a sample of the species in question, a minimum of 30 animals, mark, and then release them back into the wild. Allow the marked individuals a period of time to mix with the total population and then take a second sample of about the same number, recording the total numbers of individuals caught (say 30) and the number of these which are marked, say 5. It then follows that the ratio of the total number of individuals which were marked (30) to the total population (unknown) should be the same as the ratio between marked individuals recaptured (5) to total recapture (30).

$$\frac{\text{total no. of individuals marked and released (say 30)}}{\text{total population (unknown)}} = \frac{\text{total no. of marked individuals recaptured (say 5)}}{\text{total no. recaptured (say 30)}}$$

By simple algebra it follows that:

$$\text{total population (180)} = \frac{\text{total no. of marked (30)} \times \text{total no. recaptured (30)}}{\text{no. of marked individuals recaptured (5)}}$$

Obviously only certain types of animals can be marked with any degree of success. Fish are often marked with tags on their fins but the techniques and study of their abundance are beyond the scope of this book. Snails can be marked successfully using coloured nail varnish. Remove the snail from the water and quickly dry an area of the shell. Rub this area vigorously to remove any algae growing there. When it is completely dry, put a small streak of nail varnish on it, let this dry and return the snail to the area of water in which you caught it.

Dragonfly nymphs and some species of water bugs and beetles can also be marked. When marking dragonflies be careful to dry them quickly and mark them in the middle of the dorsal surface of an abdominal segment, not over a joint. If you varnish over a joint, you will decrease the animal's mobility and may hinder moulting. With beetles, be careful to varnish on the elytra or wing cases, as this will cause no damage to the insect. With bugs such as the water scorpion, mark the surface of the wings. Snails are quite easy to mark since they will retreat into their shells while you quickly dry and mark them but insects may present more of a problem. Dragonfly nymphs can be held gently between two fingers, although a certain amount of dexterity may be needed to prevent them from wriggling free. Water beetles are more tolerant of being out of water but require greater caution when handling since some species can bite. It is best to press them flat on a piece of card with fingers on either side of their wing cases while you mark them. Water scorpions can also be dealt with in this manner but you should handle them more gently because their bodies are softer.

It is essential that the animals are not harmed by the marking process – either by being out of the water for too long or by the mark itself. This not only avoids damage for their sake, but also, if they are damaged, you may be less likely to recapture them which will affect the result of the experiment.

Another requirement of this technique is that the marked individuals should mix randomly with the rest of the population. Do not release all the marked animals in one spot, but instead try to let them go in the area where you caught them. Otherwise, when you sample for the second time, you may get a biased result by catching too many or too few marked individuals. When recapturing, do not selectively try to catch marked individuals, but instead sample the whole area randomly. It is essential that you use the same method for capturing and recapturing.

The other factor to be considered is the time between marking and recapturing. It must be long enough for the marked individuals to mix with the others, for example two weeks with snails. In the case of dragonfly nymphs, however, it must not be too long, as they may moult and so lose the mark. In this case a period of one week may be the optimum. Time gaps will vary according to species and you may only discover the best duration through trial and error. If you want to be certain that the mark will remain for the period of the experiment, you could mark a few individuals prior to your main mark-recapture experiment and keep them under observation in an aquarium. You would then be able to decide the maximum time that could elapse between marking and recapture.

Mark-recapture is a more specialised and accurate method for studying animal abundance than a simple arbitrary scale, as described on page 119. It enables you to estimate fairly accurately the total population of a particular species in your pond rather than just its relative abundance. One drawback is that it can only be used with certain types of animal – those that can be marked easily.

The technique could be particularly useful in making a comparison between a chalky pond and a heathland pool. Snails, dragonfly nymphs and beetles could all be marked. The former would be common in chalky ponds whilst the latter two species would be more abundant in heathland areas. Mark-recapture could also be used to follow seasonal variations in certain species. The great pond snail would be a good subject since vast numbers of young appear in early summer. Fewer and fewer survive as the months pass, but those that do will grow to a considerable size. If you perform mark-recapture at different times of year then use different coloured nail varnish so that there is no possibility of confusion with a previous experiment. You could also estimate the average length and breadth of the snails and thus follow their growth throughout the year.

Water chemistry

Unfortunately, without the aid of a laboratory, there is not much that the amateur naturalist can do toward studying water chemistry. Basic observation of the soil type will give a clue to the amount and types of salts present in the water. For example, in chalky areas, the water will have a high calcium content. In contrast, heathland water is likely to be deficient in nutrients and acidic. The degree of acidity or alkalinity of water is measured on a pH scale from one to 14 which is a measurement of dissolved hydrogen ions in the water. A pH value of 7.0 means the water is neutral; below that it is acidic and above it is alkaline. Although the pH can change according to the time of year, it is still a useful pointer and can be easily estimated using universal indicator paper which changes colour according to pH, red denoting acids and blue alkalis. It is normally supplied with a colour chart with which you can compare your test paper.

Having determined the acidity or alkalinity of the water, you should then have more idea of the type of organisms which it will contain. For example, water with a pH of 4 to 5 would be acidic and you could be fairly certain that no snails would be present, since their shells would start to dissolve. However, this type of water is often good for dragonfly nymphs.

Water temperature

The other freshwater parameter that can be measured is temperature. An ordinary thermometer can be used to compare temperatures during the day. With care, you could also compare surface temperature just under the surface. However, be careful to read the thermometer with the bulb still at the depth you are examining. To do this either tilt the thermometer or hold it horizontally. A better alternative is to use an immersible maximum/minimum thermometer, obtainable from gardening shops, where you can read the temperature after it has been removed from the water. In this way you can obtain the temperatures of the surface layer and the bottom of the pond; if this is deep, lower the thermometer on a string. During a summer's day the surface layer may be warmer than the bottom. However, during the winter when ice is forming on the surface, the reverse will be found. By doing this you will be able to demonstrate how the temperature in a pond is not uniform, as discussed in Chapter 1.

A year in the life of a pond

If you are able to study a local pond, it would be interesting to follow a year in its life. You could employ some of the detailed methods I have described earlier to estimate the abundance and distribution of its inhabitants and hence maximum growth rates and periods of activity. Alternatively, you could simply keep a diary of events and note down such things as the first flowering date for each plant and when you saw your first frogspawn or adult dragonfly in the year.

Each pond will differ slightly from the next, in terms of the species found there and their relative abundances, but the underlying seasonal changes will affect them all. The months of January and February will be quiet times with ice forming over the surface of the water during harsh spells. Conditions are not ideal for pond dipping because of the cold water. Many animals will be in a torpid state or hibernating, because of the low temperatures, and so you should avoid causing too much disturbance. If the weather is really severe, then water birds often congregate in large numbers on lakes or reservoirs where some of the surface is ice-free, and this can be worth going to see.

By March or April most of these birds will have returned to their breeding grounds, and on many gravel pits you will be able to watch the courtship displays of great crested grebes. Catkins of

Figure 75 In February most pond life is inactive or hibernating. A layer of ice may form over the surface during cold spells and this should never be disturbed as it protects the animals beneath.

alder and later, sallow will start to appear around the water's edge, and together with frogspawn, these are the first signs of spring in the pond. March and April will also see the hibernating animals

Figure 76 By July the pond will be transformed by luxuriant plant growth. Larval pond insects are particularly abundant at this time.

Figure 77 By October most of the plants will have died back and the animals will be starting to look for hibernation sites. Water beetles and bugs will still be present.

becoming active once more and water beetles and the water spider can be seen.

By the beginning of May many of the aquatic insects will have become more active. Dragonfly nymphs will be much in evidence, some of them preparing for emergence. The hairy dragonfly is an early emerging species and the adult can be seen during this month in certain ponds and canals. As the name implies, it is a good time to see mayflies emerging and vast swarms can sometimes be seen on calm May evenings.

If your pond had frogspawn in it earlier in the year, then May is the time to start watching for the young froglets leaving the water. Their emergence from the water coincides with the appearance of many more terrestrial insects which provide them with a source of food. Many of these insects will be feeding or living on the waterside plants which start to grow rapidly from now on. Plants such as *Phragmites* and yellow flag provide ideal cover for the many birds that nest around the pond's margin.

In June the flowers are revealed in their true glory with water lilies, brooklime, starworts and many others in abundance. Dragonflies start emerging in greater numbers and you will be able to watch their territorial battles. These continue throughout July

Figure 78 This great pond snail has an egg rope on its shell which was laid by another snail of the same species.

and are well worth studying. Different flowers appear in July with meadowsweet, water speedwells and many of the umbellifers being found in chalky areas and sundews and bog asphodel in heathland. July is also a good month for water bugs and for planktonic life and the weather is usually conducive to pond hunters!

By August and September, many of the aquatic and emergent plants will have stopped flowering and have started to die back. As I mentioned earlier, some of the surface loving plants form almost complete carpets across the surface by the end of the summer. Species such as duckweed, frogbit and water fern often combine to smother a small pond. In heathland areas many plants flower later than in more nutrient rich waters, and it is during September in particular that the delicate flowers of bladderwort, bog orchid and butterworts can be seen.

By the late summer there may be vast numbers of juvenile snails in the water. In some species the adults will have died off and the only individuals remaining will be young ones. Most of the water beetles will have developed into adults by the beginning of October and they will be easy to find until the first cold weather of November. In particularly mild autumns adult dragonflies can be found on the wing well into October and on several occasions I have seen adult sympetrums around in the middle of November. In the water itself small nymphs will have hatched from eggs laid that year. This only happens in certain species of dragonfly because in many species the egg does not hatch until the following spring.

By the time December comes most of the plants will have died back and many of the animals will be dead or in hibernation. On the bottom of the pond you will be able to find ostracods scurrying around and feeding on the wealth of dead organic material. Apart from these and a few aquatic insects, there will be relatively little activity. You will have witnessed the cycle of events that follow the seasons in our ponds, and finally it will have come to an end. But as a consolation, it will only be a short while before everything is active again.

Comparing types of pond

Once you have compiled a set of records for your local pond, it might be interesting to compare it in detail with another freshwater habitat in a different area. If your local study area is on chalk, you could study a heathland pool. An alternative would be to compare a small pond with the edge of a large lake.

Take, for example, the comparison between a heathland pool and a pond on chalk. The first thing you may notice is the difference

---

Figure 79 The behaviour of three-spined sticklebacks is fascinating to watch. These males are in their breeding colours and fighting over territory.

in vegetation. The heathland pool will be surrounded by clumps of *Juncus* and purple moor-grass. Close to the edge will be found *Sphagnum* moss and perhaps the insectivorous plant sundew. Depending upon the size of the pool, there may be submerged plants such as the pondweeds (*Potamogeton* spp.). However, in small pools, areas of *Sphagnum*, either alive or dead, may be the only vegetation. In the chalk pond, you will immediately observe much more vegetation. Watercress and fool's watercress are typical emergent plants, having their roots under water. In the water itself there will be a whole host of plants, such as Canadian pondweed, frogbit and water starwort. Maps of plant distribution and transects are particularly useful in making this type of comparison.

When comparing the fauna of two different areas, it is important, once again, to standardise your sampling methods. If sampling animals in the open water, make the same number of sweeps of your net in each area. Similarly, if you are sifting through the mud, compare the same number of scoops. You will immediately notice striking differences in the animals found in the heathland and chalk areas. Where for example, are all the snails in the heathland pond? The answer is related to the water chemistry, and in particular the calcium content of the water as has been discussed more fully in Chapter 1. In general terms, however, the heathland pond will be much less productive than the chalk pond.

As an interesting continuation of this project, you could compare how well the same plant species grows in water from the two sites. You will have observed that there may be a marked variation in the extent and type of vegetation from the two habitats and you can assume that there are basic differences in their water chemistry. One way of testing the effect of the water content is to measure plant growth. This would be rather difficult with most aquatic plants but is relatively easy with duckweed, the small oval, single-leaved pondweed which floats on the surface and can sometimes completely cover some water bodies. Its main method of reproduction is simple division and, under favourable conditions, this can occur every day. Since this division can only occur after growth has taken place, the number of new leaves that appear each day gives an estimate of growth.

Set up two identical buckets, one with heathland pond water and the other with water from the duckweed pond and place 20 duckweed leaves in each. Leave the containers in the same place with some direct light and count the number of leaves produced in each over the following weeks. You should find that division in the original pond water far exceeds that in the heathland water. You could perhaps set up a third bucket containing heathland water with a handful of chalk thrown in which will counteract some of the acidity of the water and may encourage plant growth.

Figure 80 Duckweed can often carpet the surface of a pond by the end of summer. Since its main means of reproduction is by simple division it makes a good subject for examining the difference between acidic heathland water which supports little plant life and water from chalk areas which is much more productive.

This is only one example of the effect of the properties of water and its dissolved chemicals on plant growth. Demonstrating such effects is often a good substitute for the amateur naturalist for chemical analysis of water, which is beyond his scope.

Colonisation

When a new area of freshwater is created, it is colonised by aquatic animals and plants in a relatively short time. However, not all the organisms arrive at once, nor do they colonise at the same rate. By using some of the mapping techniques described previously, you can trace the colonisation and succession as it occurs.

One of the most suitable areas for study is a newly formed gravel pit. If there are small bays, you can treat each of these in the same way as you would a pond and draw distribution maps based on a grid system. Initially there will be few plants present, but a visit every month should prove rewarding. It is important to use the grid in the same place each time, so leave four marker stakes in the ground to mark the four corners of the square grid.

If the pit is large or inaccessible in some way, then a transect may be more suitable. Again, it should be performed in the same place each month and you can record the plant species present along the same line. Transects show the overall area occupied by a plant species, but do not give any indication of the number of individuals of that species present. By performing a transect each month, remember that you will not only be looking at colonisation, but also at the growth of plants already there. It might be better to compare the transect at the same time in consecutive years. This will give a more accurate impression of the rate at which colonisation is progressing.

Animals will also be colonising the new water body and it can be fascinating to follow how long this takes and see which species arrive first. One simple way of recording animal colonisation is by noting the presence or absence of particular species. If the number of species present is plotted on a graph against time, it will rise and eventually, after several years, level off. Obviously, the first animals to arrive will be the most active. Since insects are the only invertebrates to have mastered the art of flying, they are generally the first to appear. In the first year of the gravel pit's existence dragonflies, water beetles and bugs will arrive. In subsequent years these will be followed by animals which are more confined to water, such as crustaceans. The eggs of these animals may be carried by water birds who find the new habitat more attractive as the plant and animal communities develop.

More detailed studies of colonisation will involve estimating the abundance of the animals. The easiest way to do this is by creating an arbitrary scale as described on page 119. You must ensure that your sampling technique is the same each time you visit the area. Otherwise you can influence the apparent abundance and mask any changes in numbers. Mark-recapture techniques are not as suitable in the study of colonisation because continual immigration of animals will dilute the proportion of your marked individuals to the total population. (In a mature pond the numbers of a species emigrating to colonise new areas should be roughly the same as the number arriving from other habitats.) If you particularly want to use mark-recapture techniques you could overcome this problem by having the recapture date close to the marking date. However, this may result in the marked individuals not mixing thoroughly with the rest of the population, which is essential if the experiment is to give reliable results.

Another area that could provide interesting information is a boating lake. These are frequently created as an amenity in towns and cities. Boating lakes are usually lined with concrete, and at first will contain little life. Gradually an algal growth will develop on the concrete and crustaceans such as freshwater shrimps will begin to colonise it. Sometimes boating lakes can support large populations of sticklebacks, but this will not occur until a thriving plant and animal community has developed. There is little point in trying to map the plant distribution or perform transects since the concrete lining can support very little. Animal populations and colonisation can, however, be studied by estimating the relative abundances over a period of months.

Studying the colonisation of a gravel pit or boating lake is a worthwhile project since it demonstrates how easily freshwater organisms can disperse to new areas. If you follow the colonisation of a gravel pit over a period of years, you will be able to see how succession of plant communities takes place from the truly aquatic plants to the stable scrubland of freshwater margins. As an alternative project you could follow the succession of the plant communities in a pond as it declines and finally fills in. One disadvantage with this is that the process may take a long time, or rather, over a period of a few years there will be less change than in a new area which is being colonised.

Photography in the field and at home

By far the best way of recording the life in your local pond is to photograph it. In the past many freshwater naturalists made extensive collections of 'pickled' specimens stored in alcohol. Apart from the fact that these quickly lose their colour, it seems perverse that the enjoyment of a living creature should require its death. The real interest in any form of life, including that in the pond, lies in a study of it in its natural surroundings. With a bit of care and patience you can examine and record pond life on film. You will then have a permanent record that will not lose its colour, and has not harmed the creatures at all. Tank photography is also a fascinating pastime. Again, it allows you to watch and photograph animals behaving naturally – almost as if you were in the pond with them.

The best camera to buy for general freshwater photography is undoubtedly a 35 mm single lens reflex. Most current models have through-the-lens metering (TTL) which means that you get the correct light exposure whilst looking through the camera at your subject. There are many makes on the market and a great range of prices too! Some of the more reliable are Nikon, Olympus, Pentax, Canon and Praktica. SLR cameras are excellent to use in the field because they are small, compact and easy to handle. Larger format cameras such as Hassleblads and Bronicas produce very good results but are invariably bulkier and more expensive. Another drawback with these is that of the magnification of the image, for it is in close-up work that the real rewards of freshwater photography lie. The larger the format, the longer the lens you will need to produce the same image size relative to the film. Alternatively, the larger the format, the closer you must be to your subject with the equivalent lens, and this is not always possible in the field.

For photography in the field, when close-up work is necessary, then a macro lens is desirable. This is a lens that at its closest focus produces a 1:1 image size (life size) on the film. Alternatively, there are many macro-zoom lenses on the market which will enable you to vary your distance from the subject and still produce the same size image. When photographing adult dragonflies it is a definite advantage to have one of these lenses. These insects have regular perches where they sun themselves and keep a look out for prey and intruders of the same species. However, their good vision will also enable them to see you approaching with a camera. It is very difficult to get near enough to one in the field with a 50 mm macro lens. A zoom lens that covers the range 80 to 210 mm will allow you to be further away and still get the image size that you want.

If you have taken up photography, it is best to start off by using black-and-white film. This has the advantage of being cheaper than colour film and it is also a lot cheaper to process yourself, should you wish to do so. Many people use both kinds, but colour can do more to illustrate the beauty of natural organisms. I prefer to take colour transparencies rather than prints because not only can these be used for projection, but good prints can also be made from them. To my mind the best film is Kodachrome 64 because of its sharp definition and excellent colours. It is only available at present in 35 mm format, but two other good films, Ektachrome and Agfachrome, come in a variety of formats to suit different cameras. Many people prefer these films to Kodachrome but I find the grainy appearance of the slides offsets any advantages of colour rendition.

Under many circumstances a tripod is essential. It enables you to keep the camera steady and in one position for a long period of time and alleviates the problem of camera shake. The heavier the tripod, the more stable it is. However, not all heavy tripods go low enough for the purposes of nature photography and so a compromise may have to be made.

The last piece of photographic equipment I would like to mention is the filter. There are a whole range of different types and manufacturers but in general, an ultraviolet or skylight filter will help cut out glare and also protect the front of your lens. If you are interested in photographing aquatic animals or plants from above the surface of the water, a polarising filter is essential. This will cut out the reflections from the surface if used properly and will enable you to photograph things like a trout in a fast-flowing stream.

The other considerations in obtaining a photograph of a particular subject are concerned with your own approach and skill, both of which will improve with experience. The time of day can affect not only your subject's behaviour, but also your own ability to take the photograph. In bright light you will be able to use either a fast shutter speed or a greater f number (and consequently have a smaller aperture and hence a greater depth of field). You must make the choice between these two alternatives according to the requirements of each particular photograph. If your subject is fast moving, you might choose to sacrifice depth of field for a fast shutter speed. If it is stationary, the depth of field could be given priority. Poor light conditions often create problems for photography because both shutter speed and depth of field will be reduced compared to a sunny day. Such light conditions are, however, often associated with cold weather which can have its advantages as many surface animals will be slower moving and therefore easier to photograph. If it is essential to photograph in poor light, you could consider using one or two flashguns. The best combination is that of two flashguns mounted on brackets on either side of the camera. If one flashgun is significantly brighter than the other, soft shadows will be produced giving a natural look to the photograph.

Flight photography of insects is rather too specialised for this book since it involves the use of complicated flashguns and triggers which are expensive as well. However, under bright light conditions you could attempt a few subjects. Dragonflies often hover in the same place and with care and patience you can work out the most likely spot and pre-focus on this point. It is obviously rather a hit and miss affair and you should be prepared for a lot of failures. You should use a shutter speed of at least 1/250th second or preferably 1/500th; a zoom lens would be an advantage. Mayfly swarms can also be good subjects, but they are present only for a short time; agrion damselflies stay for longer periods and the males often form swarms of up to a dozen individuals which fly up and down a particular stretch of river.

The composition of a photograph really depends on personal taste and the difficulties of the particular situation, but there are certain guidelines which can be followed. When using a macro lens, the depth of field can be rather limited so be sure that you are getting the most important part of your subject in focus and that you are taking the photograph from the best angle. If you were photographing a resting damselfly, you would get more of the animal in focus if you were perfectly side on to it. When photographing a frog, make sure that the head is in focus because this is the thing that you will look at most in the final photograph.

The position of your subject in the picture is also important. It is often a good idea for the main subject to be fairly central. Bear in mind the limits of your depth of field, because it may be better to have a relatively small subject image which is all in focus than to fill the frame with it and have most of it out of focus. (The closer you are to your subject the less depth of field you will have.) If you are aiming to isolate your subject, then try to have an out of focus background. Twigs and leaves in the background can often distract from the subject so you could do a bit of 'gardening' before you take the photograph, if the subject will permit this. If, on the other hand, you are trying to illustrate camouflage, then get as much of the surrounding vegetation in focus as you can. A black background, such as one produced by flash, can often look unnatural. The only circumstances in which it can be really effective is when the subject has a delicate structure, such as the wing of a dragonfly, where the detail might be lost against a natural background.

The fundamental key to all successful wildlife photography is patience. If you are impatient your subject will either move away from you or never come close in the first place. Less obviously, many photographs that one sees today show the animal in a state of alarm quite unlike its normal relaxed state. This should be avoided at all costs and photography without disturbance should be your aim.

Photography also has its disappointments. I remember vividly a hot afternoon in July when I found a nymph of the broad-bodied libellula dragonfly climbing up a stem of *Juncus* about to emerge. It was in an isolated pond in the middle of wooded heathland and the *Juncus* was about 1½ m from the pond's edge. I positioned myself in what I considered to be the best place for photography which meant that I was crouching with the water level over the top of my wellingtons and my trousers were gradually getting soaked. The skin of the nymph duly split open and the adult wrenched its head and thorax from the old skin and rested with its body dangling down. As is characteristic of many dragonfly species, it rested for nearly 20 minutes, during which time I did not dare move, for fear of disturbing the *Juncus*. It then continued its emergence, extricated itself from the nymphal skin and proceeded to pump up its wings, a lengthy process taking up to 20 minutes. During this time I had taken a full sequence of the emergence. Just before the wings were fully formed, a gust of wind appeared from nowhere, knocking the dragonfly off its perch and carrying it into the middle of the pond, far beyond my reach. For the dragonfly this was disastrous but I suppose it must be a fairly common cause of death. For the photographer, it was more than a little frustrating! I had a sequence of dragonfly emergence in full sunlight, but lacked the final, clinching photograph.

A more specialised approach to freshwater photography is to take the specimens back home with you. You should not do this unless you are willing to look after them. Having your specimens in the studio will allow you to take greater care over the photography and, in particular, the lighting.

Now comes the problem of what to photograph your animals in. For a lot of my freshwater photography, I have constructed a series of special photographic tanks, mentioned in Chapter 3. The dimensions of these can be whatever you choose, but in most of mine I have a front dimension of about 20 x 20 cm. The width is the critical dimension and my tanks vary from 1 to 6 cm. This allows you to restrict the movement of the subject and means that you can focus easily. In all photographs the background is important. Blue or green card placed behind the tank provides a realistic effect. Be careful not to have the card too close, because any imperfections or creases will give an obviously artificial effect. On some occasions it is desirable to have a black background and black velvet draped behind the tank is the best way to achieve this effect. However, do not use a dark field for everything. It is particularly useful though if you wish to highlight a very delicate animal or a finely divided plant.

Try not to have your photographic tank too cluttered with weed as you will find that your animal invariably chooses to hide behind it! It is also essential to use clean water, so filter pond water through muslin to remove any particles of dead plant material which would be highlighted in your photographs if you did not remove them. Avoid using tap water since this will contain chlorine used to sterilise the water and it may kill your subjects.

The next consideration must be the source of lighting. The brighter the light entering the camera, the smaller the aperture of the

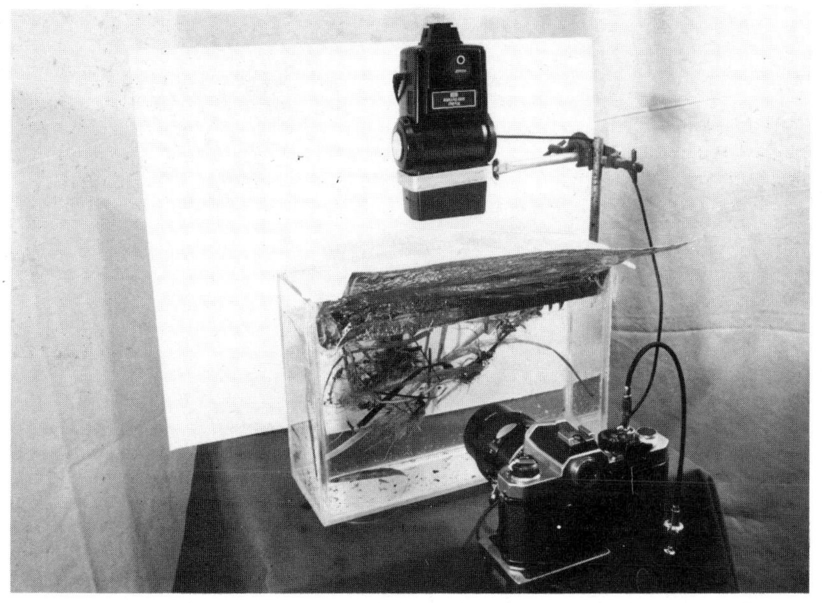

Figure 81 Here, the equipment needed to take photographs in an aquarium is assembled in position. The camera is supported on a tripod and it has a 50mm macro lens and a cable release. The flashgun is supported by a retort stand and the aquarium has aluminium foil attached to it to deflect light away from the camera body. A sheet of diffusing plastic is on top of the aquarium and the board behind provides a coloured background.

lens required, and hence the greater the depth of field. Associated with light, however, is heat. Excess heat is something that must be avoided at all costs, since by warming the water, you not only reduce its oxygen content, but also increase the animal's requirements. A flashgun can be a definite advantage over natural light or bulbs, since it gives a short period of very intense light, sufficient for your photographic requirements, but not enough to warm the water. Since the duration of the flash may be as little as 1/1000th of a second, the flash has the added advantage of 'freezing' any movement of the animal. The distance between the flash and the subject can be varied for more or less light intensity. Having placed your animal in the tank and allowed it to settle down, you must then position your flashgun. The simplest way is to suspend it a couple of centimetres above the tank using a retort stand. Use a diffuser with it if you have one, but otherwise you can place a thin piece of white plastic such as a plastic bag over the top of the tank as shown in figure 81. This helps to 'soften' the otherwise harsh flash light. The water and the reflective glass surfaces of the tank will also soften the lighting.

It is very helpful to have your camera on a tripod, since you may have to wait for some time for your subject to assume the required pose. Also, it is quite important to have the film plane (the back of the camera) as parallel to the front of the glass as possible. Otherwise, you may end up with multiple images of the subject reflected off the inside surfaces of the tank.

The light from the flashgun may also illuminate your camera and a reflection of it may appear on your film if you are not careful. You can overcome this by taping a strip of aluminium foil to the top of the front of the tank to shield the camera from the flashgun.

As far as the aperture settings are concerned, I am afraid that this will require trial and error. I use a small flashgun and with this placed 5 cm above the tank, I use settings of f 8, 11, 16 and 22, depending upon the magnification at which I am working. You could try taking a few shots on black-and-white film first before embarking on colour photography. Remember to take notes on the settings, distances and magnifications that you use. By doing this you will learn from your mistakes and will be able to predict the approximate settings you will require. Also note down the species that you have photographed so that you can label your slides accordingly; this also applies to photography in the field. Do not leave your animals in the tank for too long, since they will be starved of oxygen. Photography in small tanks is a superb way to isolate small freshwater invertebrates, but please do not use fish or amphibians since their oxygen requirements are far greater.

Obviously you can photograph whichever animals and plants you choose, but a few suggestions related to further projects may be helpful. It is always nice to take a series of photographs with a theme. For example, you could take close-up pictures of the mouthparts of various animals such as water bugs, dragonfly nymphs and water beetles. It is amazing how much detail can be seen in a slide that you miss with the naked eye.

Animals such as crayfish and stone loaches are perhaps best photographed in a tank from above. For this you can use an enamel tray with some gravel or sand in it. You can use natural light but failing this, direct two flashguns at low angles at opposite ends of the tray. This prevents reflections from the water's surface from reaching the camera. With only one flashgun you would end up with very uneven lighting and heavy shadows.

Many aquatic animals camouflage themselves in their natural surroundings and a series of pictures depicting this would be very worthwhile. Damselfly nymphs are often camouflaged amongst pondweeds for protection, whilst water scorpions lurk there waiting for prey. Nymphs of the broad-bodied libellula half-bury themselves in the mud and are extremely difficult to spot. In your photograph your aim should be to show how well the camouflage works and not to isolate the subject from its environment.

Pond skaters are good subjects for photography in the field. They do not like large objects moving around them, so position yourself and your camera in the most suitable place and wait. Keep your eyes open for struggling insects which will attract the attention of the

Figure 82. This bullhead is extremely well camouflaged against the stony bottom of a chalk stream. Your photographs should try, like this one, to show the advantages of camouflage in the natural environment.

pond skaters. You can even coax these towards you and the camera but it is rather unethical to drown an animal deliberately for the purposes of photography!

The movement of aquatic animals is fascinating to watch but rather difficult to show on a single photograph. However you could try taking a series of pictures of leeches 'looping the loop'.

Later on in the book I describe how to watch and study tadpole development and it would be a useful complement to take accompanying pictures of the various stages. By having a set of photographs you could continue your study long after the tadpoles have developed. Likewise, the courtship of newts and sticklebacks could be recorded on film which would enable you to refer back to aspects of their behaviour all year round.

The use of the microscope

The myriad microscopic plants and animals present in freshwater, although seldom seen, are vital to the successful operation of aquatic ecology. In suitable habitats they are often present in astounding numbers. However, because of their small size, they are beyond the scope of most freshwater naturalists since their study requires the specialised use of a microscope.

Few people are fortunate enough to possess their own new compound microscope because these are generally very expensive. However, most schools have a few microscopes with which you can study freshwater life and another alternative is to keep your eyes open for a secondhand instrument. A few shops specialise in secondhand scientific equipment (see Appendix III) and occasionally ordinary secondhand shops or house sales can turn one up. There are a large number of makes available including both monocular (with one eyepiece) and binocular (two eyepiece) models. There are certain points to look for that both types should have in common. The microscope should have a heavy enough base for it to be stable. Also, if it has been well cared for, the glass lenses should be unscratched and the focusing smooth and easy. Although not essential, a variety of objective lenses with different magnifications are a great help.

In order to obtain the maximum benefit from a microscope it is important to have everything set up properly. One of the most essential considerations is to have it positioned at a suitable height so that your neck and eyes are not strained. Also, give yourself enough bench space to rest your arms. These two factors can greatly improve your enjoyment.

In this section I want only to describe projects that can be carried out on living specimens without harming them. If you wish to look at mounted animals and plants I suggest you purchase these from one of the suppliers mentioned in Appendix III. Lighting is very critical both for yourself and your subject. It can heat up the slide rapidly and, unless due care is taken, can kill the organism being studied. Daylight is an acceptable means of lighting the microscope and is of a low enough intensity to do little harm. However the light is variable and often not sufficiently intense, so you may have to resort to using a bench lamp. A 40 to 50 W bulb at about 15 to 30 cm from the microscope will suffice if there is no internal light source in the microscope. Shield the direct light from your eyes as this can cause eyestrain. At the base of the compound microscope is a mirror, plane on one side and concave on the other; above this a condenser focuses the light on the slide. When using low power magnification the condenser will probably not be necessary and the plane side of the mirror can be used. At higher power the condenser will be required in conjunction with the concave side of the mirror which must be carefully adjusted. Firstly, using a high magnification, adjust the position of the concave mirror until the light from the bulb is brightest. Place a microscope slide in position and focus on this. Then bring the condenser to its highest position. Next you should mark a piece of Sellotape with a pen and stick this on the bulb. Then move the condenser until the mark comes into focus with the slide and you will have the optimum position. The Sellotape can then be removed and the microscope is ready for use. The intensity of light required will vary from slide to slide and it can be adjusted using the iris diaphragm. Your microscope should need little attention but when not in use, keep it covered to prevent dust from settling. If dust does accumulate remove it with a blow-brush and clean the lens with lens tissue.

The first project that you are likely to carry out with a microscope is the observation and identification of microscopic organisms from your pond. It may not be possible in all cases to make a full identification, since this requires specialist knowledge. You will, however, be able to identify in most cases the families to which the animals and plants belong. Make a note of where in the pond your samples came from and keep them in separate labelled bottles. A drop of water can then be transferred to a microscope slide using a pipette and covered with a cover slip. You will find differences in the organisms from the water column and from the mud in the same pond. Animals such as *Amoeba* prefer the bottom sediments whilst most planktonic algae will be found in the water column.

The water column can be best sampled with a plankton net as described in Chapter 3. If you are lucky you may come across the large alga *Volvox* which forms a colonial sphere. Your water column samples and those from the bottom sediments will also contain dead material. Keep an eye open for fragments from large plants and pollen grains as well as the remains of small insects, especially wing cases.

Having become familiar with the microscopic life in your local pond, you could try comparing it with water from a different area. As well as noting the different species present, you could estimate their relative abundances by counting the numbers in the same volume of water.

Many protozoa feed by simple ingestion or engulfing by the cell wall. This can be observed in species such as *Amoeba* by a process called vital staining. The stain, which demonstrates living processes, should do little harm to the animals. It can be purchased from suppliers (see Appendix III). You will need a thick culture of *Amoeba* which can be grown on a lettuce diet. The lettuce should be dried in a low oven and then powdered and mixed with pond water. Within a week or so you will have a dense enough culture to perform the experiment. With most animals a drop of water containing the organism should be placed in the middle of the slide and the cover slip gently lowered on top. Support this with two pieces of hair to prevent the animals from being squashed. With larger organisms, cavity slides can be bought but these will not be necessary with protozoa. With *Amoeba* a dye called Neutral Red should be used at about 1:10,000 dilution. A couple of drops of the dye in 100 ml of the culture should be adequate to indicate pH. The food vacuoles within the *Amoeba* can be watched changing from red to orange to yellow as the pH varies from acid to alkali. Finally, when digestion is complete, the contents of the vacuole will be shed into the water.

Congo Red indicator can be used with *Paramecium* cultures. First stain some milk with the dye for a few hours. The fat in the milk will stain red and, when added to the culture, will be taken up into food vacuoles. Here the dye turns blue due to acid secretion and through purple to red again as alkaline conditions are restored after digestion is complete. If the organisms are too active to be easily visible you can try restricting their movement by placing filamentous algae under the cover slip.

Dragonfly emergence

One of the wonders of insect life is the process called metamorphosis. This literally means changing body and is used to describe the emergence of a butterfly from a chrysalis for example. Many insects such as bugs only undergo partial or incomplete metamorphosis. Here the nymphal stages are like small versions of the adult. A good example of a complete metamorphosis that occurs in freshwater is that of the dragonfly. Here the larval or nymphal stage looks nothing like the adult insect and the metamorphosis is a dramatic change.

The emergence of dragonflies from their nymphs can be readily watched at home. To do this, collect a few nymphs early in the year, but remember to do it before April or you may well upset some that are about to emerge. These can be kept at home in tanks stocked with animals such as water slaters or freshwater shrimps as a food supply. The nymphs should not be crowded and should preferably have their own small tanks. One last addition to the tanks should be a stout twig and some dead reeds. If the twig is long enough it can be propped up on the side of the tank, but failing that it can be secured vertically in a plasticine block on the bottom. It should be long enough for about ½ m to rise above the water and will be the stem up which the nymph crawls to emerge.

It is not until May or June that things start to happen. For a few days before it is due to emerge, the nymph may attempt to crawl up various stems in the tank at dusk. By doing this it locates the best site for emergence and also gives you an advance warning of when it intends to emerge. The tank should not be disturbed in any way until the dragonfly has emerged one or two days later since this may upset the nymphs. In the penultimate day before emergence the nymph skin may start to lose its brown appearance. The colours of the dragonfly may be seen through the skin of the abdomen and the colours of the compound eyes will show up. During this last stage in

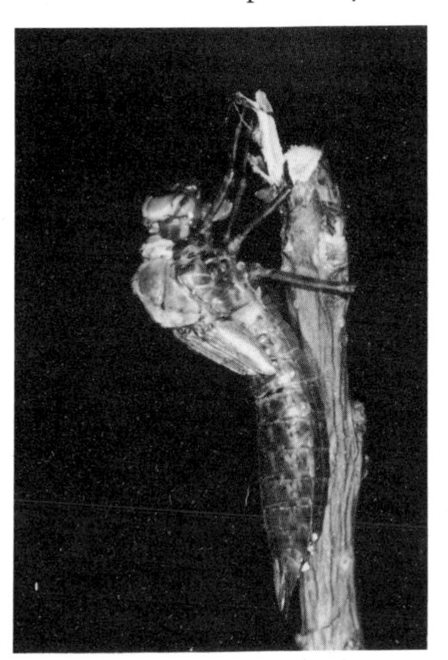

Figure 83 The first stages of dragonfly emergence. The nymph climbs up a twig at dusk and after a resting period the skin splits down the back of the thorax. The head and thorax of the adult can be seen (left). Eventually the head, thorax and part of the abdomen are pulled free from the nymphal skin (right). The adult hangs motionless for some time before continuing to emerge. The white threads are the linings to the tracheae or air tubes.

Figure 84 After its second resting period the adult pulls itself free from the nymphal skin (left). The body is soft and the wings are crumpled. The wings are pumped up over a period of half an hour and it is imperative that the dragonfly should not be disturbed at this time. When the wings are fully expanded (right) they harden. After a few hours the dragonfly takes its maiden flight. The true colours do not develop for several days.

the nymph's life it will not feed. The 'mask' that was used to catch prey will have lost its muscular attachments and have become redundant.

When you can see adult colours through the nymphal skin, you should be prepared to watch the emergence. This happens at night in many species to avoid excessive heat and predators and so you will have to make special preparations. Do not switch on lights during the emergence process as this will cause the nymphs to retreat back into the water. If emergence has started and a light is switched on, the nymph will try and move to avoid the light and this could have fatal consequences. It is better to have soft lighting already on when the sun sets and the nymph should not be put off by this.

The first step in emergence is for the nymph to crawl up the chosen twig or stem. Having gained a firm grasp, the skin of the

Figure 85 This smooth newt larva is excellently camouflaged amongst pond weeds. It still has its external gills and well developed legs.
Figure 86 After eight weeks this tadpole has fully developed hind legs whilst its left front leg is pushing through the operculum slit. In contrast to the newt larva it has lost its external gills.

thorax and posterior part of the head will split and you will see rhythmic pumping movements in the body as the adult forces itself out. Eventually the whole head, thorax, legs and the first part of the abdomen will have been withdrawn from the old skin. The soft body of the adult will then dangle helplessly in a resting phase which may last 10 or 15 minutes. You will also be able to see thin white threads attached to the inside of the nymphal skin. These are the linings to the tracheae or air tubes which have been pulled from the tracheae of the adult dragonfly. During this stage of metamorphosis it will have changed from 'breathing' dissolved oxygen in the water to breathing air. It may be very tempting to try to assist the dragonfly in its extrication from the nymphal skin but please do not. You may not only damage its soft body, but will also disrupt the sequence of metamorphosis. This will undoubtedly result in the dragonfly's death or in a crippled specimen.

After sufficient rest, the dragonfly quickly arches its body and secures a hold on the nymphal skin. Because of its firm grasp, it is able to pull the remaining segments of the abdomen free from the old skin and will then hang from this. The wings will appear crumpled and opaque at this stage, but again rhythmic pumping movements may be seen as the dragonfly inflates the 'veins' of the wing causing their expansion. This is another critical period for the dragonfly because the process of chitinisation or hardening of the external skeleton will have begun. This means that, unless the wings are quickly pumped up to their full size, they will harden in a crumpled state. So do not do anything that could cause disturbance. After about half an hour the wings will be fully formed and dry. The dragonfly will take a maiden flight after a further hour or so's rest.

If you are familiar with the coloration of the species of dragonfly that has emerged, then you will notice that the colour is very different at this stage. The adult colours will take several days, if not weeks, to appear, by which time it will be sexually mature. In order to mature, the dragonfly must fly and eat. For this it is necessary to release it in a suitable area, preferably the one from which the nymph came.

Once the dragonfly has emerged, the dried nymphal skin or exuvia will remain clinging to the reed or stem for some time. These are suitable things to make a collection of, and given a little practice, you can readily distinguish between the different species, using the key in C.O. Hammond's book mentioned in the Further reading section.

If you are an early bird, you can try to observe newly emerged

---

Figure 87 A young common frog sitting on a frogbit leaf. Having completed its metamorphosis, it will soon leave the pond where it was spawned.
Figure 88 Common toads mating in the position known as amplexus. The smaller male grips the female's arms and is carried around by her.

dragonflies or late individuals emerging in the field. This will necessitate your arriving at the pond before dawn and you can inspect suitable emergence sites such as reeds and twigs above the water's surface, by torchlight. The dried exuviae provide a good indication of the numbers of dragonflies that have emerged and this is a useful project to carry out.

On Lundy in 1978, whilst accompanying a school party on a field course, we studied the numbers of the common sympetrum dragonfly emerging from a small pond. The pond had fairly uniform vegetation around its perimeter and so we assumed that dragonfly emergence would take place all around the edge and not in localised spots. We then chose 15 random sites around the edge and marked off 1-metre lengths along the circumference. For a few days or weeks before they are due to emerge, nymphs of many dragonfly species, particularly the sympetrums, appear to congregate around the edges of the ponds in which they are living. We therefore attempted to collect all the nymphs along the 15-metre lengths, as far out into the pond as we could. By taking an average figure for the number of nymphs per metre length and multiplying this by the circumference of the pond we were able to estimate the total number of nymphs present in the pond. Our 15 samples were performed at the same time and after counting the nymphs were returned to the water. To this figure we added the number of exuviae already present around the edge of the pond. We also removed all the exuviae we could find in our 15 sample strips and counted the new ones each day. In this way we were able to estimate the rate of emergence and could also correlate it to the weather conditions on the day. The value of 1,200 for the total number of nymphs in the pond seems incredibly high for such a small area of freshwater, and gives an indication of the wealth of life in British waters.

Another project suggestion is to estimate the numbers of adult dragonflies present during the season on your pond. You could then watch for set periods of time and see how many eggs are laid. This is only feasible with species such as the libellula dragonflies which dip their abdomens onto the water to lay eggs, one at a time. By multiplying this number by the hours of daylight, and the period of the season during which the dragonflies are egg laying, you could estimate the total number of eggs laid. The egg-laying period will vary according to species and the conditions of the pond.

These figures, although only a rough estimate, might give you a valuable insight into the life history of the dragonfly species you are studying. The number of eggs laid would vastly exceed the numbers of nymphs, and this in turn would be greater than the number of adults emerging. Many of these adults would disperse from the pond after emergence and many more still would be eaten before they could lay their eggs. This illustrates why many animals that do not show parental care, such as dragonflies, lay a vast excess of eggs to ensure the survival of a few through to the adult stage as discussed in Chapter 5.

Watching tadpoles develop

There can be few people who cannot remember, as children, having collected frogspawn and watching tadpoles develop into frogs. By returning to this fascinating childhood pastime later in life you can not only learn a great deal about their development but also do the frogs a good turn. Frogs have declined rapidly over the last 10 or 20 years, probably mostly as a result of changes in agricultural practices, such as draining of wetlands and increased use of pesticides. Wet meadows with ditches and pools are havens for frogs and other pond animals, but seem to be regarded by the unenlightened as wasted land.

Frogs produce vast numbers of eggs because many will perish before they reach the adult frog stage. Perhaps as few as one in a hundred survive their development. So, considering the vast numbers of initial eggs, it will do no harm if you remove a few and look after them at home. It can be a positive benefit if you are able to bring most of them through to young froglets. The whole process takes about two months and during that time should provide a great deal of interest. Try not to take too much frogspawn, since the less you take the better you can look after the developing tadpoles. Place them in a separate tank with some pond weeds so that they are not eaten and so that you can clean it without disturbance to other organisms.

The eggs are laid in vast masses, each one coated in a jelly-like substance. This swells up on contact with water forming the familiar frogspawn and is rich in energy-giving substances. The egg inside the jelly at this stage looks like a black dot, but on closer inspection is darker above than below. The energy in the jelly enables divisions to take place in the black dot. The first one is vertical, followed closely by the second which is also vertical but at right angles to the first. The third division is horizontal and this separates four small dark segments above from four paler segments below. The former is the embryo whilst the latter will give rise to the yolk. The next two divisions are by two, giving 16 and 32 cells, but from then on the divisions are more complex and irregular. Eventually a hollow sphere called the blastula is formed. Soon the dark cells divide to cover the pale cells except for a small area and this stage is known as the gastrula.

The pale area forms a slit which deepens until the embryo is a double walled sphere with a hole at one end. The sphere then elongates with a head and tail end forming the typical comma-

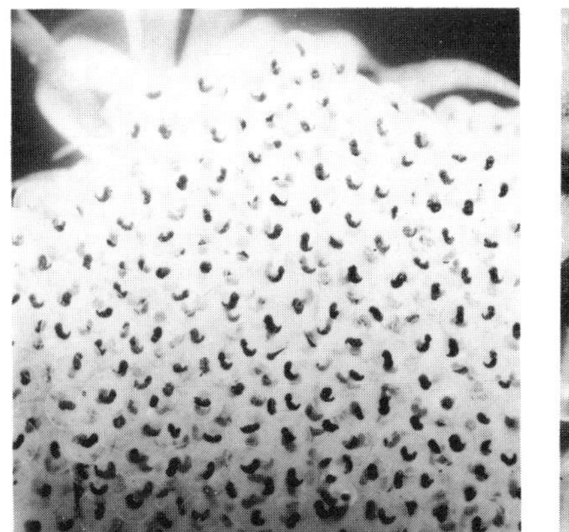

Figure 89 This sequence of four photographs shows tadpole development. Eggs are laid in vast quantities. They are protected by a jelly which forms the familiar frogspawn (left). These young tadpoles (right) are about two to three weeks old. They are herbivores at first but later become carnivores.

Figure 90 Hind legs soon develop on the tadpoles and these are then followed by the front legs which burst through the skin. Once the legs are fully formed, the tail is reabsorbed and the froglet will soon leave the pond.

shaped tadpole. Inside the embryo the organs are developing and after about two weeks the tadpole is ready to hatch.

Initially it cannot swim properly and remains attached to its jelly by the sucker and a glue-like secretion. Inside the tadpole there is still a lot of development taking place and when the mouth becomes properly connected to the gut the tadpole can move to a plant and graze algae. These are rasped off surfaces and it is essential to have plenty of plants in the tank with them.

At this stage the tadpole breathes with its feathery gills but a flap of skin called the operculum then grows over them. A single opening or spiracle remains and the tadpoles must take water into the mouth and expel it over the gill rather like fish.

At about the time that the tadpole loses its sucker and its gut becomes visible, its diet becomes at least partly carnivorous. At this stage the number of tadpoles per tank should be reduced or they will eat each other. You can easily feed them by tying a small piece of meat to some cotton and dangling this in. The meat should be changed each day as should the water to stop it becoming stagnant. The tadpoles are best removed using a tablespoon to prevent damage to their soft bodies.

Hindlegs will soon appear on the tadpoles' bodies followed by the forelimbs. One of the forelimbs grows through the skin but the other bursts through the operculum. The next stage in development is for the tail to be reabsorbed and you should now add a few freshwater invertebrates such as copepods or small freshwater shrimps or freshwater slaters. The tadpoles will now rely on their lungs rather than their gills and so stones or bricks should be placed in the water so that they can climb out. Be sure to cover the tanks from now on with muslin or they will all escape. The young froglets can be released in your garden pond or in a newly reclaimed gravel pit. In your garden pond you could record the numbers released and those that return each year and when they first start breeding.

Ostracod reproduction

Ostracods are, as you may remember, the tiny bivalved crustaceans that can be seen moving rapidly across the sediment in most, if not all, ponds and lakes. There are many species, but these are difficult to identify without a microscope, since the largest is only 2 mm long.

The project is based on some research done by Dr Peter Henderson on an ostracod called *Cypridopsis vidua*. This species is easily recognisable since it has a characteristic piebald appearance. You can collect a few from your local pond for this project, or use any other species that is plentiful. Despite their small size, ostracods can be rather difficult to collect because of the speed at which they move. The best way to catch them is with a 'wide-bore' pipette, such as an eye-dropper, and a rapid suck is usually successful. Having collected a few specimens, build up a stock culture of the ostracods in a tank with plenty of detritus. After a few weeks you should have more than enough for your needs.

In the wild, ostracods feed on rotting plants and animals and lay their eggs around their food source. It is upon this principle that the project is based, and in captivity they will readily take to agar jelly. This can be bought in most chemist's shops and should be prepared by dissolving 5 g in 100 ml water that has boiled (to remove air bubbles) and cooled to 80° C. Pour the solution into a tray and let it cool. At around 40°C it will set and can be cut into regular 1 cm squares.

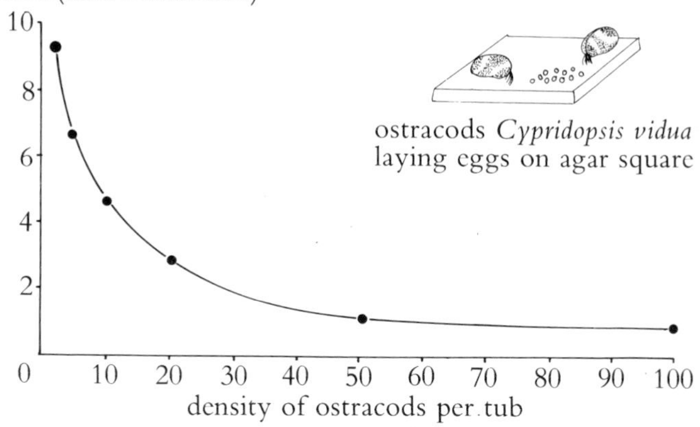

Figure 91 This graph shows clearly that the number of eggs laid by individual ostracods decreases in direct proportion to the density of ostracods present. (after Henderson)

You should then set up six identical beakers or old margarine tubs containing the same volume of water (about 10 ml) and one 1cm square of agar and place a different number of ostracods in each, covering the range one, five, ten, 20, 50 and 100 individuals per beaker. The ostracods will eat the agar and lay their small, pearly white eggs on it. These are easily visible through a magnifying glass or even with the naked eye.

Count the number of eggs produced each day in each tub and divide this number by the number of ostracods present to give you a value for the number of eggs laid per individual. You should find that your results are similar to those in the graph shown above which demonstrates that the greater the density of ostracods, the fewer eggs are laid per individual.

Why should ostracods lay fewer eggs in higher densities? In the natural environment food sources, such as fresh leaves, will be scattered over the bottom of the pond. The higher the density of ostracods around a particular food item, the more quickly it will be eaten. Consequently, fewer generations will be supported by this item of food as the density increases. For the ostracods it is better to lay fewer eggs when their density is high and wait for the opportunity to lay on a fresh piece of food. But it is a puzzle as to how the ostracods can detect differences in density. They only have simple eyes and being parthenogenetic, do not need to come together to mate. By a series of analyses Dr Henderson found that they detect levels of an enzyme called amylase in the water around them. Amylase breaks down starch to form maltose and is released by the ostracods to digest their food. The concentration in the water is thus related to the number of ostracods present, and Dr Henderson demonstrated the same effect using artificial amylase.

With a bit of luck, you should be able to see this relationship between density and egg laying at home. It shows that even these tiny animals have quite a complex relationship with the environment around them.

Smooth newt courtship

Smooth newts have a lovely courtship display which can easily be watched at home in your tank in early spring. Unlike frogs, the male newt never grasps the female in the embrace called amplexus; instead male and female perform graceful movements to ensure that fertilisation of the eggs occurs. The courtship movements of the male can be broadly categorised into wave, whip and fan movements. The wave movements consist of the male positioning himself in front of the female side on and waving his tail so that she can see his crest and the patterns on his underside. Whip movements involve a vigorous lashing of the tail to wash water over the female. Lastly, fan movements occur when the male curves his tail, touches the body of the female and vibrates his tail like a fan. The current that this causes carries odours to her.

Next, the male deposits a packet of sperm on the ground. At this moment the female lies behind the male with her nose touching his tail. The male then moves forward exactly one body length so that when the female follows, her cloaca is over the sperm packet. She then clasps the packet in her cloaca and internal fertilisation occurs.

These precise movements are necessary to ensure that the sperm packet is not lost. The female then has to lay the eggs, also a delicate operation requiring precision. When she has found a suitable leaf, she grasps it with her hind legs, lays a single egg and folds the leaf over for protection.

Watching newt courtship can provide a fascinating insight into animal behaviour. After studying them, please return both the newts and their eggs to the pond from which they came.

Figure 92 The female great crested newt lays her eggs singly and wraps each one delicately in a leaf.

Breeding behaviour of sticklebacks

As a follow-up to newt courtship, sticklebacks also make intriguing study and can be kept successfully in a tank at home. During the breeding season, the male is a blaze of colour, having a bright red belly, blue eyes and a green back. In the wild in early spring he stakes out a territory with a gravelly bottom and some weeds surrounding it. He will also set up a territory in a tank and defend it against intruders so do not be tempted to introduce a second male since they will fight continually. A male will even respond vigorously to his own image in a mirror in the tank. The red belly ellicits this response and a model with a red splash of paint will have the same effect.

The male constructs a nest in the centre of the territory. Initially, a shallow pit is created by removing mouthfuls of gravel and this is filled with algae bound together by a sticky secretion from the kidneys. He then burrows through this mixture to make a tunnel, in which the eggs will be laid.

His next task is to attract a female stickleback to his nest. At this time in the wild, the females move about in shoals and can be recognised by their silvery appearance and swollen bellies. When a shoal swims past, the male gets visibly excited. His colours become even more intense and he rushes out to meet them. He performs a zig-zag dance in front of them scaring off all but receptive females. The female of his choice is enticed back to the nest where he first demonstrates the tunnel. She then wriggles through it and at the critical moment he stimulates her to lay her eggs. The signals for this are a series of nudges and vibrations along her flanks by his head. The male then wriggles through himself and sheds sperm over the eggs to fertilise them.

After this point the male sees the female as an intruder and chases her off. He then sets about guarding and caring for the eggs. A flow of water must be maintained over them to provide oxygen. The male diligently fans water with his pectoral fins at intervals during the ten days before the eggs hatch.

The young fish are also guarded by the male who stops them straying by catching them in his mouth and spitting them out at the nest site. After about two weeks they start to evade him and eventually rush to the surface to gulp air and fill their swim

Figure 93 The courtship of sticklebacks is complex and rewarding to watch. Here, a male three-spined stickleback escorts a female to the nest he has prepared.

bladders. Thereafter they fend for themselves and many perish in the process.

Sticklebacks are fascinating to keep in an aquarium because of their vivid colouring during the breeding season and interesting behaviour. The courtship displays of the male are followed by complex and prolonged parental care which adds to their attraction. However, please release the young into a suitable water body when they have left the care of the male.

Watching molluscs feeding

Many species of mollusc make useful additions to an aquarium and aspects of their behaviour can be studied by watching the tank. Limnaeid and planorbid snails, such as the great pond snail and the ramshorn, feed at least partly by grazing algae off the surfaces of plants and stones. In their mouths they have a rasping tongue called a radula with which they scour the surfaces. In your aquarium this can be best watched on the surface of the glass and if there is a coating of algae then the snail will leave a zig-zag trail where it has been feeding.

The bivalve molluscs are completely different from freshwater snails in that their bodies are enclosed in two separate hinged shell valves which can be closed with powerful muscles. They have only limited powers of movement and live buried in the mud at the bottom of ponds and streams. In order to obtain food and oxygen they have tubes or siphons and ciliary action generates a current of water which passes over the gills and mouth. This current can be easily demonstrated by placing a swan mussel in a tray of water. Once it has settled down and opened its shell valves, the siphons will be extended. If you pipette some silt or dye towards the longer of the two siphons, it will gradually be sucked in. After a period of a few seconds, it will appear out of the exhalent siphon. Their filter feeding habits mean that bivalves accumulate toxins and certain bacteria in their bodies, seemingly with little detrimental effect. It is not a good idea to keep swan mussels in the aquarium since they frequently die in captivity and their large bodies can quickly turn the water anaerobic and kill everything else.

These are a few suggestions as to projects that can be carried out in most ponds in this country or your aquarium at home, but of course there are numerous others that you can design yourself. By recording information and making detailed observations you will get much more pleasure out of your pond dipping, and gain more knowledge as well. However, you should always bear in mind the welfare of the animals and plants that you are studying, and try to harm them, and their environment, as little as possible.

7 Conserving the freshwater environment

In Britain the freshwater environment is under threat from neglect, wilful destruction and pollution. Over the last 50 years we have lost perhaps as much as 50 per cent of our suitable habitat and to combat this, conservation of the areas remaining is essential. However, to have convincing reasons for the conservation of freshwater habitats, it is important to understand the need for conservation in a global sense.

The arguments for conservation fall into three basic categories: survival, economic and moral. It is sometimes difficult to separate the first two which hinge on Man's dependence upon his environment for his continued existence. Man requires the plants in the ecosystem both for food and for the production of oxygen necessary for respiration. When plants photosynthesise they use up carbon dioxide which is the waste product of respiration. Without plants, Man and the rest of the ecosystem would be unable to survive and so from this point of view it is essential to have a flourishing environment. It would seem sensible, therefore, to avoid unnecessary destruction of anything in the environment, but each individual's perception of what is necessary and unnecessary differs. This is at the centre of the moral argument.

Man is not dependent on plants alone. Animals too, are an integral part of many people's diet, although whether this is necessary or not is debatable. Perhaps more importantly, animals together with plants and the environment as a whole, are essential for the recycling of nutrients in the ecosystem. The freshwater environment plays a particularly important role in this recycling of nutrients and in the water cycle. The disappearance or pollution of freshwater habitats upsets the water cycle and could have serious consequences both for human health and for agriculture. If all the freshwater habitats in Britain were removed, we might not notice any significant effects in the short term, but in the long run we certainly would. In this sense, freshwater habitats in Britain are analagous to tropical rainforests, the destruction of which is causing previously undreamt of consequences through soil erosion and climatic alteration.

Destruction of the environment for short-term gains is short-sighted in another respect. Many plants at present are sources of medicine and food and many new uses for them are being discovered all the time. Plant geneticists are able to improve yields and pest resistance in existing crops by careful selective breeding. In future years, at present untried plant species may yield strains which could produce valuable crops.

As I suggested earlier, it is harder to convince people with a moral argument. Man does, however, have a responsibility to stop the destruction of wildlife when this does not adversely affect his own survival. We simply just do not have the right to eliminate animal or plant species for ever. Whaling products, for example, are all obtainable from other sources and the benefits of whaling are purely economic. Ironically those who have an interest in whaling can present sound economic reasons for saving whales from extinction in order that they may be exploited for a longer period of time. Campaigners against whaling do not want to see whales exploited now or at any time in the future. Here, the economic argument draws people's attention away from the moral argument of whether whaling is necessary for the survival of mankind.

Finally, the reasons for the preservation of both habitat and species should not simply be based on the economic or survival benefits they provide for mankind. There is the aesthetic appeal and educational benefit of a complete environment. Ponds, and therefore the environment in general, should be preserved, not only for the present generation, but also for future ones – *to enjoy*. Perhaps it is here, more than anywhere, that children can gain an understanding of ecology and the fundamental importance of conservation.

Unfortunately, conservationists are in a minority and subject to financial constraints which limit their ability to be effective. Because of these constraints they often have to create a scale of importance and decide that a particular species or habitat is more worthy of conservation than another. The Nature Conservancy Council grades sites according to their scientific interest and by so doing, protects many habitats and species either special to Britain, or where Britain is an important area in their world distribution. This is probably a wiser approach than that of simply conserving all rare species in Britain, irrespective of whether they are common elsewhere in their range.

Some conservationists are concerned solely with the preservation of endangered species. While this is laudable, there is an implication here that Man can continue to exploit such species as long as they are not endangered. Some people apparently see nothing wrong in this, but I would have thought that any sincere conservationist would also be concerned to some extent with animal welfare and the unnecessary destruction of any wildlife. Conservationists can also easily be lulled into a false sense of security by the existence of nature reserves. However, the land that they occupy is only a minute fraction of the total land area of Britain and we must be aware of this fact.

Active conservation also entails some management which can involve the removal of species from a particular area in order to preserve others and judgements must be made here. Most freshwater habitats would eventually fill in and become stable scrubland, were they not periodically cleared. It would be acceptable to allow the natural process of succession if freshwater habitats were created at the same rate as they are disappearing due to natural causes. However, they are disappearing at an alarming rate due to destruction by Man and those remaining must be actively managed if this valuable habitat is to survive.

Freshwater does far more than simply harbour aquatic animals and plants. It acts like a magnet for many other forms of life and so a great deal of wildlife suffers when ponds and marshes disappear. The natural process of succession is the easiest threat to combat in most cases since it requires a relatively small amount of periodic work. Ponds are often neglected for long periods of time, and they then require a good deal of work to restore them, as will be discussed later on. Conservationists must always be ready to point out the value of freshwater systems not only to wildlife but also in the recycling of nutrients since this is often forgotten by planners and developers. Marshes are often regarded as waste lands ripe for developing and drainage is a frequent fate.

Wetland areas are also exploited for recreational purposes. Gravel pits are seemingly ideal places for boating, water-skiing and fishing. These interests need not necessarily conflict with the interests of wildlife. The most productive areas in a large gravel pit are likely to

Figure 94 Through careful management this gravel pit at Papercourt in Surrey has become a haven for wildlife, notably Canadian geese and gulls.

be in sheltered bays and around the shores. If sufficient numbers of these are left undisturbed, then all interests could be catered for. Fishing and shooting do cause a certain amount of pollution from lead shot, apart from the destruction that results from the direct pursuit of these pastimes. The lead not only boosts dissolved lead concentrations in the water but lead shot is also readily ingested by water birds such as swans and ducks. The shot is about the same size as the particles of grit that these birds swallow to help in digestion of food and so they have little chance of avoiding it. Once inside them, the lead causes muscle weakening and in swans the drooping neck is, unfortunately, an all too common sight. A recent NCC survey suggests that around 3,000 swans die each year from lead poisoning, and perhaps an alternative to lead shot and weights is the only solution to this problem.

Pollution is nowadays a major threat to our freshwater habitats. Since ponds became redundant as watering places for livestock, they have often been turned into rubbish dumps for the local population. Although unsightly, these do little to harm the environment, except perhaps clog it up and speed up its decline. Organic pollution is far more insidious. The human population in this country has increased tremendously over the last 200 years and, apart from the sheer physical burden of all these people, the removal of their sewage has been, and still is, a tremendous problem. Luckily, these days sewage treatment processes have advanced and very little raw sewage enters our rivers. However, although treated sewage has none of the problems of disease associated with it, it still has a very high content of organic matter.

The organisms largely responsible for the breakdown of organic material are bacteria. Although some bacteria can function perfectly well without oxygen, many use it to respire. If there is a high

Figure 95 Unfortunately, unsightly cases of rubbish dumping in ponds still occur. This may not harm the pond but it can speed up its decline.

content of organic matter in the water, the bacteria will soon build up, and deplete the oxygen which inevitably affects other animals in the water. Thus, only those animals which are specially adapted to low oxygen concentrations or which breathe aerial oxygen will be able to survive.

Another form of organic pollution is the run-off from agricultural fertilisers which leach through the soil into rivers and ponds. These enrich the nutrient levels of water and enable algae, especially blue-greens to flourish. The result is an algal 'bloom' and the end result where the oxygen in the water is depleted, is known as eutrophication. Plants, as you will remember, respire oxygen as well as producing it by photosynthesis. Not only do the blue-green algae use up a lot of oxygen, they also cloud the water preventing light from penetrating to the submerged plants.

Chemical pollution can have the same effects as pollution by agricultural fertilisers. Pesticides applied to crops can affect many other animals apart from the target organism. Such was the case with DDT which was used against insect pests before it was discovered that accumulation of the chemical occurred in each step of the food web. Thus, the higher up the food chain, the greater the accumulation and the more lethal the effects. In the terrestrial environment the effects were first noticed amongst predators such as birds of prey. Not only were these found to be dying but they had a lower breeding success due, amongst other things, to egg-shell thinning. In the freshwater environment it was the fish in particular that suffered. Despite widespread concern it still took some time for the use of this chemical to become banned – such was the short-sighted nature of some of the farming community.

Heavy metals such as lead, copper and arsenic are also highly toxic even at levels down to parts per million. They are toxic to all animals and in the freshwater environment there is no escape for most organisms. Their levels are strictly monitored by water authorities and industries producing them as waste products appear to show some degree of concern, as far as the law goes. However even a slight spillage can result in massive 'kills' in the freshwater environment.

Such drastic effects are relatively easy to spot. If you go to a water body that you know was full of life and suddenly it is lifeless, the cause may be chemical. In this case you should immediately inform your local water authority. But what of the more insidious forms of organic pollution and how can these be detected?

Earlier in the book I discussed the various adaptations of freshwater organisms for gaining oxygen and showed that some animals are better equipped for survival in low oxygen concentrations than others. Thus we can create a scale with animals tolerant and intolerant of organic pollution. By studying your local pond or stream you can deduce the level of pollution by the types of animals that are found there. I have included streams in this section as they are often more prone to pollution, being sites of drainage. In streams, stoneflies, mayflies and blackflies are least tolerant to depletion of oxygen as a result of organic pollution and their presence will indicate an unpolluted area. In an unpolluted pond there should be some species of mayflies, but the important groups to look out for are dragonflies and beetle larvae.

The freshwater shrimp and freshwater louse which may be found in flowing and still waters can tolerate a certain degree of pollution. In streams, the next down the list are net-spinning caddis larvae and chironomids. The latter will be found in still water along with leeches which can be grouped together on the scale.

Now we come to the organisms which are most tolerant of pollution and are true indicators of a polluted water body when found on their own. The *Tubifex* worms are typical indicators but are not as tolerant as some fly larvae such as the rat-tailed maggot. Mosquito larvae often thrive in waters that are polluted to some degree at least, partly because their predators are unable to survive. However they are filter feeders and must have a supply of planktonic algae. *Tubifex* worms and the rat-tailed maggot are detritivores and can therefore live in isolation.

You can use this pollution scale to check the level of pollution in your local pond or stream. If it is obviously polluted then you should inform your local water authority. If it is in a healthy state then keep a regular check on it. By so doing you will be performing a valuable service since you may be able to spot pollution occurring and stop it.

Another form of pollution which may not be immediately obvious is that of heat. Power stations are the main source of thermal pollution since water is used to cool the turbines and the temperature of the water used may be raised by as much as 10°C. This can obviously have a devastating effect upon animals with a strict oxygen requirement such as mayflies. However, since power stations are invariably situated on sources of flowing water, the mixing of hot and cold waters reduces the effect. In fact the effect is often unnoticeable a few kilometres down stream. Some species may even benefit from water heating. One such example is the rudd and you can often see large shoals at certain distances from the outflow of warmed water where the temperature is at an optimum for them. This is not to say, of course that the freshwater environment as a whole benefits from the heating effects of power stations; it clearly does not.

You may also encounter two forms of 'natural' pollution. In heathland areas it is not uncommon to observe small areas of water with a film of oil on the surface. This is due to the breakdown of organic material by bacteria. Because of the acidic content of the water, the bacterial fauna is limited and this 'oil' and methane gas are the end result. This is entirely natural and nothing to be alarmed about. In similar habitats you may also see areas of what look like 'rust' by the side of streams and in marshy places. This too is natural and a result of the work of iron-fixing bacteria.

What you can do

Perhaps the most immediate thing that you can do, apart from monitoring your local pond for pollution, is to help in or organise, the clearance of your local pond or canal if it has become overgrown or derelict. This is best done as a member of a group, but before you attempt anything you must be sure that the pond really does need clearing and not just managing. It is essential that you get permission from your local council or the pond owner, and seek the advice of your county naturalist's trust which will give valuable encouragement. Some councils are quite keen to encourage groups of naturalists to conserve their local environment and may even provide assistance. You could also try to get the area designated a nature reserve after you have finished. The Royal Society for Nature Conservation and its junior body, the Watch Club are useful organisations to join since they keep you in touch with what is happening nationally as well as locally. There is also an organisation called the British Trust for Conservation Volunteers which specialises in conservation projects including pond clearance. They cannot help in all cases since they are short-staffed but you could help in another way by joining them and assisting in various weekend tasks. (See Appendix III for addresses.)

If you decide to clear your local pond, then your first task is to do a survey beforehand to discover what, if any, plants and animals are

Figure 96 It helps if you can take an interest in the welfare of your local pond and also encourage others to do so.

present and which are scarce and common. In past years recolonisation from adjacent ponds could have occurred and even scarce species could have returned naturally. However we have lost so much freshwater habitat in recent years that natural recolonisation is now likely to take a longer period of time. Indeed you may even have to replant common species of aquatic vegetation if the pond has decayed too far, or otherwise you will end up with a pond full of unwanted blue-green algae and nothing else. Rare or scarce species should be collected before clearance takes place and they can then be returned when the pond has settled down. In this way you can avoid the problem of conservation causing more harm than good.

It is important before starting pond clearance to have a well thought-out plan and to stick to it. Clearance may require extensive work if the pond has been left for a long period of time. It may have been covered with emergent vegetation such as *Phragmites* reed or have been filled with rubbish or fallen leaves. The latter can be harmful, turning a pond anaerobic. However do not be tempted to remove any trees since a certain amount of leaf material encourages a thriving decay cycle in the water. Also trees increase the value of ponds to terrestrial animals immensely and should be regarded as part of the freshwater ecosystem albeit one that must be managed. Together with a reed margin, trees also serve as shelter breaks from the wind.

The growth of a dense reed bed, which enhances the pond's value to wildlife, requires shallow shelving sides to the pond. Indeed these are necessary for proper zonation of aquatic plants to occur. Steep sides are most unproductive and will inevitably reduce the natural value of the pond. So if the pond you are clearing has steep sides, these could be dug away during clearance.

If the pond in question still has a wealth of freshwater life present, but is in danger of filling in in the near future, then the best approach is to clear the pond in sections each year so that the cleared part of the pond can recover and be colonised from the uncleared parts before they in turn are dealt with. If you clear the whole pond in one year, you run the risk of destroying many of the plant and animal species present. In some ponds, most of the original life may have disappeared already and you will have to reintroduce it. Be careful to introduce only those plants that would have occurred naturally and keep notes and records on how they fare.

The operation should ideally take place in the autumn since at this time of year least harm will be caused to breeding species. Indeed many plants will have died back and some animals such as frogs and toads will have left the water for the winter. Unless you have a large team, successful clearance is only possible in a fairly small water body, and the first stage is to remove any large logs and branches together with rubbish such as prams and tyres. The bottom of the pond can be dredged with a grappling iron to remove excess weed and branches, or if you have a boat, an old rake would serve. The weed should be left on the bank to allow animals to crawl back into the water. If a large amount of mud has collected at the bottom of the pond then a drag line attached to a tractor or a mechanical excavator (if you can get hold of one!) can be used. As I mentioned before, try not to be too vigorous in your clearance since you want the pond to recover eventually.

At first, the cleared part of the pond will look rather unsightly, but by the spring new plant growth will have occurred and it will soon recover. Clearance is not the end of the story since, as I discussed in earlier chapters, the process of succession will continue. You therefore need a management plan and may have to perform

Figure 97 Restoration of ponds and canals can have a dramatic effect, transforming a dumping ground into a haven for wildlife. Before restoration work began on the Basingstoke Canal many parts of it has become overgrown and clogged with rubbish.

Figure 98 This photograph shows the same stretch of canal after restoration. The lock gates are now functional and the freshwater habitat has been improved immensely.

small-scale clearance operations every three or four years. If silting has been a problem, then this too will continue after clearance. You may have to consider putting a silt trap in the feed stream just before it enters the pond. This is a concrete trench, sunk below the stream bed where silt can collect and be dug out easily.

Canals can suffer the same fate as ponds if left untended. It is currently popular to restore canals for their amenity value and this has happened on the Basingstoke canal. In this particular instance, the interests of conservationists and the canal society have to a greater degree been satisfied. By liaison, the canal was dredged in sections rather than all at once, thus reducing the devastating effect of removing all the water. After each section had been cleared, there were adjacent areas from which colonisation could take place. Had all the water been removed at once, colonisation would have taken years to occur from distant ponds. The Basingstoke canal was formerly of national importance as a site for dragonflies and rare species, such as the hairy dragonfly and the metallic emerald were formerly plentiful. Let us hope that in future years they will thrive alongside the restored canal barges. A few weeks after a stretch of the canal near Aldershot had been dredged, I visited the site, and to my delight found not only water scorpions and water stick-insects but also nymphs of the hairy dragonfly – so there is some cause for optimism.

Conservation of our village ponds and canals can play a vital part in the preservation of the freshwater environment for two reasons. Firstly there are a large number of village ponds and canals and together they form a significant volume of water. Secondly they form links in the network of freshwater bodies across Britain encouraging the colonisation of new habitats. Careful, well-planned conservation of your local pond together with pollution monitoring, are valuable aids to the conservation of Britain's freshwater habitat as a whole.

Making your own pond

By making your own pond you can bring the study of freshwater life quite literally to your own doorstep and also help the conservation of aquatic organisms by providing another habitat. A pond will increase your garden's appeal to other forms of wildlife, in particular birds and mammals.

Before constructing a pond there are several considerations that should be taken into account. The size of a pond may appear to be a matter of purely personal choice. Do avoid making it too small, however, since it will be subject to temperature extremes. The larger the volume of water, the greater the buffering effect of the water on the heat from the sun and attacks of frost. Depending upon the size of your garden, a suitable size would be a 2 to 4 m diameter.

The shape is entirely up to you and the constraints of your garden, but an irregular shape has a more natural appearance.

The depth can be quite critical since a shallow pond may freeze completely. If this happens it may not only kill the pond organisms, but also crack the pond lining. A depth of about 1m is a suitable minimum and a smaller area of about ½ m depth will provide a contrast and be suitable for fish and amphibians to breed. It is important to make the sides of the pond shelve gently so that frogs and toads can leave at the end of the breeding season. This will also encourage birds and mammals to visit the pond to drink and enable any creatures that fall in to escape.

Careful consideration should be given to the siting of the pond. Obviously the ground should be as level as possible, but if there is a slope in the garden the pond could be sited at the top. This will enable excess water to drain off rather than flood over the surrounding ground if the level rises. It is not a good idea to site the pond directly under a tree, since the falling leaves in autumn will clog it up. You may also have problems digging the hole because of the roots. However trees in the vicinity will increase the aesthetic appeal of the pond and attract more bird life.

Lining the pond

The next decision should be on the material with which to construct the pond. In most areas, garden soil is too porous to retain water and so you will have to resort to artificial means. One of the best ways is to use plastic sheeting obtainable from gardening centres. The pit must be dug to the required depth and diameter. Then you must remove all the stones or cover the soil with a couple of centimetres of sand or sifted soil to prevent any sharp edges puncturing the plastic when the water is pressing it down. Be sure to use enough plastic to line the pit completely and give a ½ m or so overlap margin around the edge. This can be weighed down with stones after the pond has been filled and disguised with soil or turves.

The next step is to fill the pond in order to check for leaks. This will also allow any toxic chemicals from the plastic to leach into the water; the water can be drained off and any leaks repaired with sealing tape. The plastic sheeting should then be covered with a layer of sifted soil a few centimetres thick.

Nowadays there is a great variety of reinforced plastic and fibre glass moulds for ponds which can be used as an alternative to plastic sheeting. Obviously you are restricted to a limited range of sizes and shapes, but for the purposes of a small pond there should be something suitable. There are, however, two essential requirements for successful use. Firstly the hole must be dug to fit their shape exactly, otherwise you may stress certain parts of the mould. One way around this problem is to dig a generous hole so that the mould easily fits in and then to fill it in with sifted soil which can be shaped to fit the mould. Secondly the moulds should be soaked for up to a week before draining and refilling to allow toxic chemicals to leach out of the plastic.

Concrete is favoured by many people for lining ponds as it has the advantage of being strong and watertight. It is, however, susceptible to cracking, especially by frosts. The hole should be dug to the required dimensions as before, and again ensure the sides are gently shelving to prevent the concrete from slipping. The cement should be on the 'dry' side and mixed with sand. It is a good idea to have a couple of helpers because it is important to form a complete layer of concrete before it sets; any joins in the concrete inevitably give rise to leaks at a later date. The best time of year to make the pond is in the autumn as the weather is more likely to be cool and damp. This helps slow down the setting process, and the slower the concrete sets, the harder it binds. Concrete also leaches chemicals into the water which harm animals and plants, so the pond should be 'soaked' and drained before filling for use. Alternatively a sealant coating can be painted on over the concrete. Some people prefer to have raised concrete ponds which are certainly easier to observe but they are more prone to frost damage and require a lot more work to make. A strong brick surround should be constructed first and the approximate internal shape of the pond built up with more bricks. You may have problems lining this type of pond with concrete, and wooden boarding may have to be used to retain the concrete on the sides while it sets.

Stocking the pond

Rooted aquatic plants such as water lilies, arrowhead or watermint should be added before filling the pond. Their roots can either be buried in the soil at the bottom of the pond or they can be planted in pots. Once the pond has been filled, you can add the other floating plants. Hornworts and starworts are both very suitable plants and can be bought from most aquarist shops. It is probably better to avoid removing plants from the wild since this is illegal without the landowner's permission and, in any case, depletes the natural environment. Canadian pondweed and duckweed are also suitable but rather prone to growing rapidly and filling the pond and should be avoided unless you are willing to maintain their numbers actively.

The plants should be left for about a month before any invertebrates are added. You can of course wait for natural colonisation to occur but by stocking the pond you can speed things up. Hopefully it should not be necessary to restock the pond after the first time since, to my mind, a really good pond is one that looks after itself, apart from a little maintenance during the winter.

If you want your pond to be a miniature version of a local pond then a small selection of its invertebrates can be added. Most animals are entirely suitable but two species, the great diving beetle and the water boatman could perhaps be omitted since they are voracious carnivores. Personally I find that the interest they provide makes up for their carnivorous diet; after all, carnivores are an essential part of any ecosystem.

Amphibians such as frogs and toads are welcome additions to any pond. Both are becoming scarce in the wild so the best plan is to remove a small quantity of frogspawn from a local pond in the spring and let the tadpoles develop in your own garden pond.

Most people would consider fish an essential part of a garden pond but there is a danger that you may overstock it if you are not careful. For a pond of the dimensions described, I would not recommend any more than six to ten small fish. Members of the carp family such as goldfish or crucian carp are suitable and will 'grow into' the pond and if they breed, their numbers will be limited naturally by its size. If you do want to breed fish then it is a good idea to have a small annexe that can be partitioned off to prevent the other fish from eating all the young. However, once they are large enough to be reintroduced to the whole pond some inevitably die. This must be accepted as a natural aspect of pond life and enough will survive to replace the older fish when they die. If you want a really natural pond why not introduce a few sticklebacks? These are charming animals and their courtship and territorial displays compensate for their small size.

Figure 99 This garden pond has a modern design and has recently been completed. It will soon hold a wealth of freshwater life.

Maintenance

With luck your pond should require little in the way of maintenance and will look after itself throughout most of the year. In the autumn you should remove dead and dying vegetation such as water lilies, since if left they will build up the detritus on the bottom and, over a period of years, reduce the volume of free water above. Also, their remains will enrich the water with nutrients encouraging algal blooms in summer months and turning the water to pea-green soup. This is harmless and will eventually disappear when the nutrients are used up, but it reduces the pond's appeal since you cannot see into it! Blanket weed is also encouraged by a high nutrient content and should be removed by hand if the quantity gets unacceptable. Try to avoid using chemicals to control it, since these are seldom specific to blanket weed and, unless you get the concentration correct, they will kill other organisms.

The fungal disease *Saprolegnia* which affects fish and gives them a 'woolly' appearance can be a problem if your pond is overstocked with fish. An infected fish should be removed and placed in a bowl of water containing salt at a concentration of one teaspoon per 4.5 litres. This is rather a kill-or-cure method but, hopefully the fish will recover within a few weeks. Once the fungus appears to be dying off the fish can be placed into decreasing concentrations of salt every two or three days until finally it is back in normal freshwater.

Conclusion

In this book I have tried to outline the wealth of life in freshwater and to illustrate the benefits of studying it. Pond dipping for its own sake can be great fun, but with a definite aim in mind it becomes even more rewarding. By studying pond organisms both in their natural environment and in aquaria you can learn about their distribution, abundance and behaviour and hopefully get an overall picture of freshwater ecology. Using a camera can lend an added dimension and I hope that, in whichever way you approach your study of pond life, you will also be keen to conserve it. You can easily become active in conservation, by monitoring pollution and combating the disappearance of freshwater habitats. If enough people become motivated, then there should be plenty of freshwater life around for future generations to enjoy.

Appendix I Scientific names of plants mentioned in text

Alder *Alnus glutinosa*
Amphibious bistort *Polygonum amphibium*
Autumn lady's tresses orchid *Spiranthes spiralis*
Birch *Betula* spp.
Bladderwort *Utricularia vulgaris*
Bogbean *Menyanthes trifoliata*
Bog myrtle *Myrica gale*
Branched bur-reed *Sparganium erectum*
Brooklime *Veronica beccabunga*
Canadian pondweed *Elodea canadensis*
Common centaury *Centaurium erythraea*
Curled pondweed *Potamogeton crispus*
Duckweed *Lemna* spp.
Fine-leaved water dropwort *Oenanthe aquatica*
Floating pondweed *Potamogeton natans*
Frogbit *Hydrocharis morsus-ranae*
Greater reedmace *Typha latifolia*
Greater willowherb *Epilobium hirsutum*
Hemlock water-dropwort *Oenanthe crocata*
Hornworts *Ceratophyllum* spp.
Horsetails *Equisetum* spp.
Irises *Iris* spp.
Marsh marigold *Caltha palustris*
Meadowsweet *Filipendula ulmaria*
Opposite-leaved pondweed *Groenlandia densa*
Pondweed *Potamogeton* spp.
Purple loosestrife *Lythrum salicaria*
Purple moor-grass *Molinia caerulea*
Reed *Phragmites communis*
Reed sweet-grass *Glyceria maxima*
Rushes *Juncus* spp.
Sallow *Salix atrocinerea*
Sedge *Scirpus lacustris*
Sphagnum moss *Sphagnum* spp.
Starworts *Callitriche* spp.
Sundew *Drosera* spp.
Sweet flag *Acoras calamus*
Thread-leaved water crowfoot *Ranunculus trichophyllus*
Watercress *Nasturtium officinale*
Water crowfoot *Ranunculus aquatilis*
Water fern *Azolla* spp.
Water lily *Nymphaea alba*
Water mint *Mentha aquatica*
Water milfoils *Myriophyllum* spp.
Water parsnip (Fool's watercress) *Apium nodiflorum*
Water plantain *Alisma plantago-aquatica*
Water speedwell *Veronica anagallis-aquatica*
Water starworts *Callitriche* spp.
Wild thyme *Thymus serpyllum*
Yellow flag *Iris pseudacorus*
Yellow-wort *Blackstonia perfoliata*

Fungi
Goldfish fungus *Saprolegnia*★

Algae
Blue-green alga *Anabaena*★
Colonial alga *Volvox*★
Filamentous alga *Cladophora*★
Filamentous alga *Spirogyra*★
Flagellate *Euglena viridis*★
Green alga *Chlorella*★
Stoneworts *Chara* spp.
Marine alga (Gutweed) *Enteromorpha intestinalis*

★These species have no common name

Appendix II Scientific names of animals mentioned in text

Protozoa
Amoeba★
Opalina ranarum★
Paramecium★
Vorticella★

Sponges
Pond sponge *Spongilla lacustris*
River sponge *Spongilla fluviatilis*

Coelenterates
Hydra *Hydra viridissima*

'Worms'
annelid worms
Eiseniella tetrahedra★
Tubifex spp.★
Nais spp.★
Chaetogaster spp.★
Earthworm *Lumbricus terrestris*
Fish leech *Pisciola geometra*
Horse leech
Haemopsis sanguisuga
Leeches
Glossiphonia complanata★
Erpobdella octoculata★
Liver fluke *Fasciola hepatica*
Medicinal leech
Hirudo medicinalis
nematode worms
Ascaris lumbricoides
planarian worms
Dendrocoelum lacteum
Polycelis nigra
rhabdocoel turbellarian worm
Microstomum spp.★
tapeworms
Bothriocephalus claviceps★
Schistocephalus solidus★
Ligula intestinalis★

Arthropods

Crustaceans
copepod *Cyclops* spp.★
Crayfish *Astacus pallipes*
crustacean *Leptodora kindtii*★
Triops cancriformis★
Fairy shrimp
Chirocephalus diaphanus
Fish louse *Argulus foliaceus*
Freshwater shrimp
Gammarus pulex
ostracod *Cypridopsis vidua*★
parasitic copepod
Ergasilus gibbus★
Water flea *Daphnia* spp.
Water louse *Asellus aquaticus*

Insects
alderfly *Sialis lutaria*★
Banded agrion *Agrion splendens*
Black sympetrum
Sympetrum danae
Blue-tailed damselfly
Ishnura elegans
Broad-bodied libellula
Libellula depressa
Caddis flies *Triaenodes* spp.★
Limnephilus spp.★
China mark moth *Nymphula* spp.
Demoiselle agrion *Agrion virgo*
Downy emerald *Cordulia aenia*
Drone fly
Eristalis (Tubifera) tenax
Emperor *Anax imperator*
Great diving beetle
Dytiscus marginalis
'Greendrake' mayfly
Ephemera danica
Groundhopper *Tetrix undulata*
Lesser water boatman
Corixa punctata
midge *Chironomus* spp.★
Mosquito *Culex* spp.
Norfolk aeshna dragonfly
Aeshna isosceles
Pond olive mayfly
Cloëon dipterum
Pond skater *Gerris* spp.
Rat-tailed maggot *Eristalis tenax*
Saucer bug *Ilyocoris cimicoides*
Screech beetle *Hygrobia nermanni*
Silver diving beetle
Hydrous piceus
Southern aeshna dragonfly
Aeshna cyanea

Springtail *Podura aquatica*
Stonefly *Isoperla* spp.
Water beetle *Acilius sulcatus*
Water beetle *Hydrobius fuscipes*
Water boatman *Notonecta glauca*
Water bugs *Plea leachii*★
Aphelocheirus montandoni★
Water cricket *Velia caprai*
Water measurer *Hydrometra stagnorum*
Water scorpion *Nepa cinerea*
Water stick-insect *Ranatra linearis*

Spiders
Swamp spider *Dolomedes fimbriatus*
Water mite *Hydrarachna* spp.
Water spider *Argyroneta aquatica*

Snails
Common bithynia *Bithynia tentaculata*
Dwarf pond snail *Limnaea truncatula*
Freshwater winkle *Viviparus* spp.
Great pond snail *Limnaea stagnalis*
Jenkin's spire shell *Potamopyrgus (Hydrobia) jenkinsi*
Lake limpet *Ancylus lacustris*
Orb shell *Sphaerium* spp.
Pea shell *Pisidium* spp.
Ramshorn *Planorbis planorbis*
River limpet *Ancylastrum fluviatile*
Swan mussel *Anodonta cygnea*
Wandering snail *Limnaea peregra*

Fish
Carp *Cyprinus carpio*
Eel *Anguilla anguilla*
Gudgeon *Gobio gobio*
Lamprey *Lampetra* spp.
Perch *Perca fluviatilis*
Pike *Esox lucius*
Roach *Rutilus rutilus*
Rudd *Scardinius erythrophtalmus*
Stone loach *Noemacheilus barbatulus*
Tench *Tinca tinca*
Three-spined stickleback *Gasterosteus aculeatus*

Amphibia
Common frog *Rana temporaria*
Common toad *Bufo bufo*
Edible frog *Rana esculenta*
Great crested newt *Triturus cristatus*
Marsh frog *Rana ridibunda*
Natterjack toad *Bufo calaminta*
Palmate newt *Triturus helveticus*
Smooth newt *Triturus vulgaris*
★These species have no common name

Appendix III Useful addresses

Field Studies Council, The Information Office, Preston Montford Field Centre, Montford Bridge, Shrewsbury. The FSC have a number of field centres around the country which run courses on many subjects including freshwater biology. The centres at Slapton Ley in Devon, Malham Tarn in Yorkshire and Flatford Mill in Suffolk are particularly geared towards the aquatic environment and details of their courses can be obtained from the above address.

Freshwater Biological Association, The Ferry House, Far Sawrey, Ambleside, Westmorland. Members receive an annual report and can use the laboratories and library. The FBA publishes a number of identification guides to the trickier groups of freshwater animals which are available to non-members for a small sum. They can also supply specimens and collecting equipment.

Philip Harris, Oldmixon, Weston-Super-Mare, Avon. This firm can supply you with most of your pond dipping equipment, including microscope accessories such as slides and cover slips.

The Royal Society for Nature Conservation, The Green, Nettleham, Lincolnshire, LN2 2NR. The RSNC is the central body concerned with County Naturalists' Trusts and can put you in touch with your local branch. By becoming a member you can become involved in local natural history and will receive a copy of the RSNC magazine *Natural World* which gives details of local and national news.

The Watch Club Address as above. The junior club of the RSNC which works in close association with the County Naturalists' Trusts. There are frequent outings, some of which will be concerned with freshwater life.

Further reading

Identification guides

Bagenal, T.B. *The Observer's Book of Freshwater Fishes.* Warne, 1941. Rev. ed. 1970.
Brightman, F.H. and Nicholson, B.E. *The Oxford book of flowerless plants.* Oxford University Press, 1974.
Chinery, M. *A Field Guide to the Insects of Britain and Northern Europe.* Collins, 1973.
Clegg, J. *The Observer's Book of Pond Life.* Warne, 1956.
Freshwater Life of the British Isles. Wayside and Woodland Series. Warne, 1965.
Colyer, C.N. and Hammond, C.O. *Flies of the British Isles.* Warne, 1968.
Engelhardt, W. *The Young Specialist looks at Pond life.* Burke, 1964.
Fitter, R. and Blamey, M. *Wild flowers of Britain and Northern Europe.* Collins, 1964.
Hammond, C.O. *The Dragonflies of Great Britain and Ireland.* Curwen, 1977.
Haslam, S.M., Sinker, C.A. and Wolseley, P.A. *British Water Plants.* Field Studies Council, 1975.
Linssen, E.F. *Beetles of the British Isles.* Wayside and Woodland Series (2 vols). Warne, 1959.
Longfield, C. *The Dragonflies of the British Isles.* Wayside and Woodland Series. Warne, 1949.
Martin, W. Keble. *The Concise British Flora in Colour.* Michael Joseph and Ebury Press, 1969.
Mellanby, H. *Animal Life in Freshwater.* Methuen, 1963.
McMillan, N.F. *British Shells.* Warne, 1973.
Muus, B.J. and Dahlstrom, P. *Collin's Guide to the Freshwater Fishes of Britain and Europe.* Collins, 1971.
Quigley, M. *Invertebrates of Streams and Rivers – a key to identification.* Edward Arnold, 1977.
Southwood, T.R.E. and Leston, D. *Land and Water bugs of the British Isles.* Wayside and Woodland Series. Warne, 1959.

Identification guides published by the Freshwater Biological Association

5. *A Key to the British Species of Freshwater Cladocera*. D.J. Scourfield and J.P. Harding.
8. *Keys to the British Species of Aquatic Megaloptera and Neuroptera*. D.E. Kimmins.
13. *A Key to the British Fresh- and Brackish-Water Gastropods*. T.T. Macan.
14. *A Key to the British Freshwater Leeches*. K.H. Mann.
15. *A Revised Key to the Adults of British Species of Ephemeroptera*. D.E. Kimmins.
16. *A Revised Key to the British Water Bugs (Hemiptera-Heteroptera)*. T.T. Macan.
17. *A Key to the Adults and Nymphs of the British Stoneflies (Plecoptera)*. H.B.N. Hynes.
18. *A Key to the British Freshwater Cyclopid and Calanoid Copepods*. J.P. Harding and W.A. Smith.
19. *A Key to the British Species of Crustacea: Malacostraca occurring in Freshwater*. T.T. Macan, H.B.N. Hynes and W.D. Williams.
20. *A Key to the Nymphs of British Species of Ephemeroptera*. T.T. Macan.
22. *A Guide for the Identification of British Aquatic Oligochaeta*. R.O. Brinkhurst.
23. *A Key to the British Species of Freshwater Triclads*. T.B. Reynoldson.
24. *A Key to the British Species of Simuliidae (Diptera) in the Larval, Pupal and Adult Stages*. Lewis Davies.
27. *A Key to the British Freshwater Fishes*. P.S. Maitland.

General

Brown, A. Leadley. *Ecology of Freshwater*. Heinemann, 1971.
Burton, R. *Ponds their wildlife and upkeep*. David & Charles, 1977.
Clark, E. *Fieldwork in Biology – an environmental approach*. Macmillan, 1973.
Corbet, P.S., Longfield, C. and Moore, N.W. *Dragonflies*. New Naturalist Series No. 4. Collins, 1960.
Frost, W.E. and Brown, M.E. *The Trout*. New Naturalist Series No. 21. Collins, 1967.
Harris, J.R. *An Angler's Entomology*. New Naturalist Series No. 24. Collins, 1970.
Imms, A.D. *Insect Natural History*. New Naturalist Series No. 8. Collins, 1973.
Macan, T.T. *Freshwater Ecology*. Longmans, 1963.
Ponds and Lakes. George Allen & Unwin, 1973.
Macan, T.T. and Worthington, E.B. *Life in Lakes and Rivers*. New Naturalist Series No. 15. Collins, 1951.

Index

Page references in *italic* refer to illustrations; page references in **bold** refer to recognition details in Chapter 4.

Acknowledgements

Colour photographs

Biofotos: Heather Angel 36, 125; Bruce Coleman: Jane Burton 90, 126, 143 bottom, Kim Taylor 54, 108, Roger Wilmshurst 18 top; Nature Photographers: F.V. Blackburn 17, N.A. Callow 35 top, Paul Sterry 144 top; Oxford Scientific Films: G.I. Bernard 72, 89, Peter Parks 35 bottom; Premaphotos Wildlife: K.G. Preston-Mafham 53 top; Bryan Sage 18 bottom; Paul Sterry 53 bottom, 71, 107, 143 top, 144 bottom.

Black-and-white photographs

Ardea: P. Morris 123 top and bottom, 124; Biofotos: Heather Angel 23, 29, 102, 111 bottom, 112, 163 top; S.C. Bisserot 12; Bruce Coleman: Jane Burton 21 bottom, 148 top left and right, Eric Crichton 25; Halcyon: Bob Gibbons 28, 32 bottom, 161: Frank W. Lane: Harald Doering Ebersberg 104; Nature Photographers: F.V. Blackburn 163 bottom, Andrew Cleave 19 top, 24, 26, 27, Janet Corbett 32 top, J.V. and G.R. Harrison 22, 167, D.F.E. Russell 19 bottom, David Sewell 157, Paul Sterry 103, 141 left and right, 142 left and right; Oxford Scientific Films: G.I. Bernard 148 centre and bottom, 152, Dr. J.A.L. Cooke 111 top, John Paling 101, David Thompson 153; Paul Sterry 21 top, 30, 39, 42, 98, 106, 135, 137; Wildlife Picture Agency: John Beach 109, Leslie Jackman 129, Rodger Jackman 158.